DARK STARS

An astrological guide to the unseen
foci in the horoscope.

DARK STARS

Invisible Focal Points in Astrology

by

Bernard Fitzwalter
Raymond Henry

THE AQUARIAN PRESS

First published 1988

British Library Cataloguing in Publication Data

Fitzwalter, Bernard
Dark stars.
1. Astrology   2. Stars
I. Title   II. Henry, Raymond
133.5   BF1671

ISBN 0-85030-643-4

The Aquarian Press is part of the Thorsons Publishing Group, Wellingborough, Northamptonshire, NN8 2RQ, England

Printed in Great Britain by
Woolnough Bookbinding, Irthlingborough, Northamptonshire

1 3 5 7 9 10 8 6 4 2

To Boris Kamenenko
without whom . . .

CONTENTS

TABLES

APPENDICES

INTRODUCTION

Modern astrology is a funny business. On the one hand, it recognizes that all the energy in the cosmos is balanced, and that for every movement in a given direction there is a corresponding opposite movement; on the other hand, it seems to have forgotten that this must also be true of the energies generated by the planets themselves.

All the planets orbit the Sun along elliptical paths, and ellipses, as we learned at school, have two focal points. It would not be too hard to calculate the location of the other focus, and project that onto the zodiac for astrological interpretation.

We are not the first people to have considered this 'Dark Sun'. It seems to have been very important to the Egyptians, good mathematicians as they were, and they celebrated certain times of the year when the two foci were in alignment.

It makes sense — the Earth goes around this point as surely as it goes around the Sun, and so it must be, astrologically speaking, a focus for our energies. What value should be given to it? This book explores some of the possibilities, makes some suggestions, and gives some examples.

Each of the planets is in a different orbit, describing a different ellipse. All of them have the Sun at one focus, of course, but the other focus is different in the case of each planet. As these various Dark Suns fall in the various houses of a natal horoscope, they could provide important clues as to the eventual disposition and deployment of those planetary energies in the individual. In short, an astrologer knows where the energies came from — and now he can see where they are going to.

There are some more places in the sky which should be considered, too. The nodes of the Moon are where the Moon crosses the ecliptic, and have been much used in astrology for hundreds of years. Rightly so; but all of the planets have nodes, and the fact is all but ignored. As the Moon's nodes show significant times in life, do not the planet's nodes show the energies of the planets interacting with the pattern of life? Are these not sudden, overwhelming movements which seem to rise from nowhere but influence us all? Again this book explores the possibilities, makes some suggestions, and gives examples.

There is a natural balance to the solar system; the harmony of 'circling planets, singing on their way', as the hymn puts it, is a long-cherished ideal.

There is a mute note in the harmony, though, and once more it is one which modern astrology chooses not to consider. That note is Phaethon, the planet which was broken, and whose wreckage is the asteroid belt as it is today. However, the harmony of the solar system is such that Phaethon's position can be determined, and the whole pattern of the solar system restored for the astrologer.

In the horoscope, Phaethon makes the link between the individual and his racial and genetic history. It shows the proper place of man and his scope within time. It shows what we have inherited from the past, and what legacy we leave to those who follow us. For this reason, it is significant in the horoscopes of those who fight to preserve the heritage and responsibility of mankind, such as ecologists. We venture possible interpretations of this intriguing lost planet, and offer examples, as before.

This is the start of a new phase in astrology. A hundred years ago, the astrologer was asked to answer the question 'What will happen to me?' For most of the last 30 years, the question has been 'What am I really like?', representing the viewpoint of the post-war generation seeking its identity. Now the question is 'What should I be doing with my talents?' The answer to that is best found through looking to the Dark Stars. Energy comes into the individual through the visible planets; it must be returned to the cosmos through the other foci if the circulation is to be maintained. As with the two terminals of a battery, the energy is seeking to go from one to the other. In a world where most of us have all we could wish for except purpose and direction, this 'astrology of the returning circuit' gives us those very things.

Intentions and Aims

This book is intended as a workbook. Although it has a fair amount of astrological theory in it, the main aim is to inspire and enable the user to work with these nodes and foci for himself. Accordingly, tables and other data have been given to enable any astrologer to locate them on any horoscope for a date in this century.

We have tried to show the theory behind our interpretations, in the hope that it will stimulate thought and experiment; only when many more astrologers start putting the planetary nodes and Phaethon into their horoscopes can a better idea of the significance of these points be formed.

We have also tried to encourage the development of symbolic interpretation. This is, we believe, at the heart of astrology. It is also a highly subjective process, and one which can be triggered by a variety of stimuli. We have given interpretative values suggested to us by zodiacal principles, by dates in history, and by reference to example charts. They are not intended as absolute and final interpretations, but as starting points for the imagination and intuition of astrologers' using these Dark Stars in their work.

The diagrams in this book are in the standard 'continental' wheel format.

Bernard Fitzwalter
Raymond Henry
London, Spring 1987

PART ONE: PHAETHON

1.

PHAETHON: HISTORY AND MYTHOLOGY

Raymond Henry

The history of Phaethon began to intrigue the author seriously in 1982 when he came upon mention of it in a book by two Americans who had visited the USSR to investigate the efforts of various Soviet psychics and other explorers of areas generally labelled 'occult'. A doctor was found, for example, whose patients had dreams of diseases from which they were not thought to be suffering; careful investigation revealed the diseases in their earliest forms, which would have escaped normal diagnosis, but which could be quickly and effectively treated. (See Gris & Dick, *The New Soviet Psychic Discoveries,* Prentice-Hall, New Jersey, 1978.)

Among the metal-benders and radionics experts interviewed by Gris and Dick were a group of scientists who had been investigating a band of tektite and silicate debris found extending from Soviet Western Asia right down to Australia. Tektites are small chunks of matter in various shapes and colours which appear to have arrived here from somewhere else, in much the same manner as meteorites. They seem to contain a lot of silicon, as well as aluminium oxides, iron, and traces of other elements. From their investigation of these the Soviet scientists had come to the firm conclusion that a single planet really had once existed in an orbit between those of Mars and Jupiter, where we now have the hundred thousand or so pieces of rock called the asteroid belt.

By the well-known 'Bode's Law'* there should indeed be a planet in that region, and few people have been satisfied by the bald statement in popular astronomy books that 'for some reason' the material there just failed to agglomerate into a planet. The material must have been gaseous to begin with, just as the other planetary

* See Appendix A, p. 167.

bodies were, and there is no reason at all why failure to agglomerate should have occurred. Moreover, if it had occurred, then there is even less likelihood that it would have formed small pieces of rock by the thousand. It must surely have either simply dissipated or been absorbed by Jupiter, Mars, Earth, or the Sun. Therefore we must conclude that there was once another member of our planetary family. Since it is now in a hundred thousand pieces, we must also conclude that at some time in the past it broke up — or was destroyed.

The story of this cosmic disaster crops up all over the world in sacred writings and mythologies. Hebraic, Sumerian, Hindu, Tibetan, Chinese, Egyptian, and Greek cultures all have record of it, and there are African and South American sources too. The stories have been interwoven and overlaid through the years as different stages in the various cultures make their contribution, but the general story line stays remarkably constant. There seems to be a central figure, Phaethon in the Greek version, who represents a source of light; sometimes he is the son of the Sun god. On one notable occasion he brings his light too close to the earth, so that he appears to be brighter than the Sun; he also burns the earth with his flames, bringing disaster to mankind. The ruling deity of the greater cosmos is forced to intervene, and the errant light source is destroyed. The order of the heavens has been seriously disrupted, though, and there are fires and floods to be suffered on Earth before the universe recovers itself. There is an excellent exposition of the Phaethon myths, and their astronomical meaning, in de Santillana & von Dechend's *Hamlet's Mill*.

The myth of Phaethon as being a description of a genuine celestial occurrence within human history is also examined in Clube and Napier's *Cosmic Serpent*. They, too, find their attention directed towards the asteroids, and from there towards the Apollo asteroids and the larger comets. They are in no doubt as to the truth of the story, and advance the idea that the fall of Phaethon may correspond to the impact of a substantial piece of interplanetary debris in historical times. Otto Muck reaches similar conclusions in his book on Atlantis.

The idea of there being a planet which was once whole and was later broken invites theory and speculation. The first begged question is 'What broke it up?' and after that 'If not what, who?' Having posited an intelligent force capable of breaking a planet, it is a very small step to convert them into a humanoid race, if only because

our imaginations tend to think of things that way. At this stage the
whole panoply of pseudo-science popularized by von Daniken and
others drops into place and obscures our view, and if we find him
unacceptable, we can try Velikovsky instead. Whatever we think, the
Soviet scientists took to the idea that Phaethon, as they named the
planet, was once inhabited by intelligent beings something like
ourselves. From this point on the only limiting factor is human
imagination, of course, but here are two hypotheses.

The first idea is that an invading body comes into the solar system
from outside. Venus has been given the blame by some accounts,
Velikovsky and others holding that this is why that bright lady (both
of the planets held to come from elsewhere are popularly regarded
as feminine) spins retrograde and has an almost circular orbit, while
other planets move in ellipses. After the 'close encounter' she tore
on towards the Sun and got caught in her present orbit. Ancient
moralists held that this was a Divine punishment of the Phaethonians,
who were evidently naughty people, the few survivors among them
making their escape to Earth. This journey in towards the Sun was
the real reason for the 'Fall of Man'. The 'Fallen Angels' could be
a group of Phaethonians mixing with Earthmen, while the rest of
them stayed up on the Moon (by some accounts the vehicle on which
the escape was made, and a home of the Gods in many mythologies).

The second idea omits the invading planet. Phaethon's inhabitants
let their technology get too far ahead of their sense of responsibility
and tapped the inner core energy of their planet until it fell apart.
Alternatively they had a terrible war amongst themselves, and their
weaponry destroyed the whole planet. The survivors come to Earth
in either case, as in the first scenario. Some credence to the idea of
tapping the core of the planet is given by the peculiar state of Mars,
which boasts a volcano two hundred miles high (its cone above the
atmosphere), and core-like iron debris all over the planet's surface.
Perhaps matters went just a bit further than that with Phaethon.

All of these stories have holes in them, though not such gaping
rents as in the 'failure to agglomerate' hypothesis. For one thing,
the invasion by Venus upsets the entire theory behind Bode's Law,
which otherwise works very well, except for a problem with Neptune
and Pluto. Secondly, Phaethon is nearly twice as far from the Sun
as is Mars, and it must have been too cold for life unless it had
considerable internal heating. Even then it would have been a very
dismal place with no more light than we get from the Sun on a very

wet day. Even Mars, still suspected of having once been inhabited, is too cold unless it too had much more internal heat in the past than it does now (though at least that volcano suggests that it did once have more inside it).

All in all, not very satisfactory. But very recently a new possibility has come to light, and it stands up very well to examination; Phaethon was not a planet, but a star.

At first sight that may appear to have overthrown all our previous reckoning, legend, and myth of an inhabited planet, but this is not the case. It means simply that the legendary planet was orbiting not the Sun but another star.

Legends abound in our oldest, and now our most primitive, cultures that our originators came indeed from another star system rather than from another planet in this one. Izar, otherwise Epsilon Boötes, is the one quoted by cultures around the Pacific region, while ancient African sources have a preference for Algol. Still others favour Proxima Centauri, our nearest stellar companion today. Mizar, in the handle of the Plough (in the Great Bear) also gets included by some northerly tribal cultures. The common feature of all these stars is that they are binary, two stars closely revolving around each other. Actually, Proxima Centauri is not a binary star, but its very close neighbour (as seen from here) Alpha Centauri is. Moreover, estimates of the likely number of inhabited planets in the Universe have generally been close to the estimated number of binary star pairs, though 'guesses' might be a better term in both cases. The point is that we have a strong and uncanny notion that there is a connection between binary stars and the evolution of intelligent life — or even life at all.

Up to now it has been easy to dismiss stories of invasion from the stars just named, since they lie many light-years away from us. But if our own Sun has in the past been one of a binary pair then we have at once a concordance with the 'binaries evolve life' hypothesis and also with the legends of invasion from another star system. The other system was the binary partner of our own, and the distance to be travelled was not very far at all. Some rather mystical esoteric schools call the other star 'Maldek', a term meaning the ill-fated centre, and for them it was certainly not so far distant as whole light-years. The totality of all the legends, found in widely spread cultures which have been out of touch with each other for thousands of years, makes it difficult to dismiss the idea that they have all handed down the story from some common original event that really did happen.

The Ukrainian astronomer Boris Kamenenko has devoted much time to the study of these legends and to the work of his Soviet colleagues on the subject of Phaethon, and considers it all but essential that our life-bearing planet must at some time have been the beneficiary of a binary star system in order to have evolved its vast variety of life-forms, especially in view of increasing evidence that its binary-planet association with our moon Luna (too large to be truly thought of as a mere satellite) does not reach all that far back into history. He speaks in terms of thousands of years, or at most a few tens of thousands, rather than even one million. And varied life on Earth, including moderately intelligent humanoid life, has been here for longer than that by a considerable margin.

How big, then, might this companion star, Maldek or Phaethon, have been? Dr Kamenenko says it was quite small, practically of minimum possible size. The planet Jupiter (and Saturn too) is sometimes called a 'failed star' or a 'brown dwarf' by astronomers. It has never quite accrued enough material to become a nuclear fire proper, though its deep interior might well be a furnace on a small scale. To shine like a small Sun, Jupiter would need to be about ten times its present mass, and that would mean doubling its present diameter of 46,000 miles (73,600 km). So a body of 100,000 miles diameter, about one-tenth of the diameter of the Sun, would be enough, and would easily be fitted in at the 2.8 astronomical units distance from the Sun which the asteroid belt now occupies (1 a.u. = 92.96 million miles, Earth's distance from the Sun).

The resultant scene, says Kamenenko, would be one with two sets of orbiting planets at work, or even three sets. The Sun's inner planets, Mercury and Earth, with or without Venus, would have orbited the Sun. Phaethon (or Maldek) would have had its own retinue, probably including Mars and Luna along with other small planets that may have been destroyed or since captured by Jupiter and Saturn after Phaethon's own destruction. Pluto too may have been one of the family, catching the blast at just the right angle to have sent it on a course that almost dismissed it from the entire binary system. The true outer planets, gas giants Jupiter, Saturn, Uranus and Neptune, may well have orbited both stars, around some point between them but closer to the Sun.

The cause of Phaethon's destruction remains a mystery but the event is by no means an impossibility. Lesser companions of binary systems have been observed elsewhere to come to a sticky end and certainly a miniature like Phaethon would have been highly sensitive to any major disturbance such as an intrusion by a very large comet or a loose planet from the explosion of another binary in a not-too-distant region of outer Space. The Venus story is possible, and it no longer upsets Bode's Law, which would simply not have been applicable to a binary system, but which has determined the disposition of the planets since Phaethon was torn apart. For Kamenenko the binary hypothesis also takes care of another anomaly which has bothered him for a long time: since the Sun's planets appear to follow a rough scale of increasing size as their distance increases to that of Jupiter, then a nicely corresponding diminution of size from there on outwards, why does tiny Mars spoil the structural pattern? If Mars was originally a child of the lesser star, Phaethon, then the matter is explained very well indeed.

Further supporting evidence is to be found in the nature of the largest asteroid, Ceres. This still follows what would have been Phaethon's original path very closely, and is itself a small, nearly perfect sphere of considerable density, just as we might expect for the 'black dwarf' core remnant of a lost star of such small size.

And now another long-persistent notion may be explained: that of life on Mars. Again, reasonable or not, it has persisted for centuries and will not go away. We have already said that Mars as we know it now is too cold for any serious life to be there but clearly this would not have been the case when it was a planet of Phaethon. It would have had internal heat of its own, as Earth has, and it would have received light from both its own parent and from uncle Sun, which even now manages to soften Mars' polar caps a little in what passes for summer there. It could have been very hospitable indeed, and could have had an adequate atmosphere too, before the great disaster (literally 'unfortunate star') blew it away, or before its own volcanic catastrophe used up all the oxygen to turn the spilled-out iron core into a desert of rust.

2.

PHAETHON IN ASTROLOGY

Bernard Fitzwalter

It is no easy thing to add a new planet to astrological practice. Although astrologers are getting quite used to such things, having had three new planets to deal with since the discovery of Uranus in the eighteenth century, and then the object known as Chiron in the last decade, much debate and experiment is necessary before the new body can become an accepted part of the astrologer's celestial vocabulary.

In the first place, the mythological implications of the name have to be considered. The naming of things is of vital importance, but astrologers are rarely in a position to do the naming themselves. Happily, new objects tend to receive names which suit their astrological identities quite well; this must say something about the mechanics of fate, or about the way astrologers think. Pluto was named after Mickey Mouse's dog, in the first instance, and *not* after the god of the underworld.

Assuming that the name is right, or at least seems to fit, and is sufficiently stimulating to the imaginative and intuitive processes on which astrological interpretation depends, then a symbol or glyph must be decided upon. Then the name, the mythology, and the glyph must somehow be blended into a set of qualities or energies associated with the planet. After that will come the determining of any possible strengths in signs, or rulership even; in doing so the corollary of rulership, detriment, must also be examined. It may even be possible to determine exaltation, something which has proved very difficult with the trans-Saturnian planets.

Phaethon comes to us from Dr Kamenenko and his colleagues already equipped with a name and a glyph. The astrologer, then, has to work with what he is given, but it is a much easier task than having to decide on a name and a glyph for himself. The facts are already before him, and all he must do is interpret them as they seem to present significance to him.

The name could hardly be bettered. If the myth of the Fall of Phaethon, and similar stories from other mythologies, really do represent the destruction of the Sun's binary partner, then to name the now-fragmented body after the central character of the myth is really the only possible course of action. Pursuing the myth and pressing it for more detail produces no anomalies, either: Phaethon is the son of the Sun, Helios, implying that the second star was of a similar nature to its father (i.e. stellar), but smaller, and under its control. There are other myths which make Phaethon one of the children of Eos, the dawn. Here he gets carried away by Aphrodite (Venus) to be the guardian of her temple. This is a very interesting myth: why did Venus interfere in the life of the Shining One (Phaethon means 'shining' or 'radiant')? Perhaps Velikovsky was right to suggest Venus as the invading planet which disrupted the solar system after all.

Phaethon drives Helios' chariot around the heavens in the myth, presumably along something not too far distant from the ecliptic; but he is unable to complete his circuit of the sky because he loses control of the chariot, and Zeus has to strike him down with a thunderbolt before the whole universe is destroyed by flame. The point in the sky at which this happens to Phaethon is suggested by Ovid, and will be worth examining later. The essence of the tale is constant in all the sources, however, and with a little imagination can even be seen in the glyph!

The glyph for Phaethon is a lower-case Greek letter phi,φ, in the style which does not have the bar extending upwards as well as downwards. It is important not to confuse it with the other style of phi, φ, which is used for the Dark Moon, Lilith.

The glyph is by no means bad, as such things go, and it is much richer in symbolic interpretations than was, for example, the 'PL' version of Pluto, ♇. It is a picture of a planet with a split through it; alternatively, especially if it is written in one stroke starting with the bowl, clockwise, it is not quite a circuit of the heavens, followed by a fall.

So much for the naming of the thing: what does it represent? Is there a new set of qualities not covered by the other planets, which this one brings to a horoscope? If the answer is *yes*, then the inference must be that previously astrology was incomplete in some way. This is a dangerous assertion, and incorrect as well, as any astrologer whose preference is for the traditional seven-planet system will be quick to affirm. Phaethon simply offers a different way of reaching the same conclusion, a confirmation of what the other planets already signify. For an idea about the *kind* of things it indicates, its own nature and history are the most reliable guides.

Phaethon is linked to a series of events long ago. Therefore it seems reasonable to link it astrologically with the idea of ancestry. Events and personalities in the past are carried forwards by successive generations into the present, and on into the future. Whatever it is that links us to our past, and that we have to offer to future generations, will be highlighted by Phaethon. An interest in previous generations and the origins of our civilization, through archaeology or palaeontology, for instance, would seem likely.

In the beginning, as has been suggested, the solar system had a different structure. Phaethon is the key to the understanding of this structure, and so astrologically it must show the way back towards the underlying, basic structure of things. This may not be, of course, the way that things are now, but it is the way that they were once, and the original layout must be taken into account if the matter is to be fully understood. Phaethon will show the roots of things, and that can be of great use in a horoscope.

Since Phaethon was once whole, and is no longer so, it serves as a reminder of the transience of all things, and also of the possibility that even the most settled and immovable objects in our experience may not last forever. If ecologists, and those who would have us pay attention to the squandering of our natural resources, would like an astrological focus for their beliefs, then Phaethon must surely be it. Linked to these notions are ideas of collective responsibility, the learning of lessons from the past, and the provision of an inheritance for future generations.

Some planets are associated with a direction; or, to be more accurate, some planets' symbolic values are better understood when a direction is used to describe them. Examples are outwards and upwards for Jupiter, and downwards for Saturn. A similar word for Phaethon may well be 'back'. Phaethon isn't backward in itself, in that it inhibits normal growth, or anything like that, nor is it retrograde (except in the usual way of outer planets, that is when opposite the Sun, geocentrically viewed); it is, however, retro*spective*, in that it is concerned primarily with the consequences of that which has gone before. It remembers the time when it was whole, so to speak, and now lives in the shadow of that remembrance. Disastrous things always happen to those who look back: Lot's wife, Eurydice . . .

The asteroids are the present reality of what was Phaethon. They are many, they are fragmented, they are dispersed. Astrological Phaethon must therefore represent unity, singularity, concentration. It is the core of things, and that provides as good a keyword as any to apply in interpreting Phaethon's position in a horoscope. The position of Phaethon in the horoscope of David Ben-Gurion, for instance, as given later, gives a very good idea of the 'core' of his existence — what he was about, what he really represented.

Astrological Phaethon, as placed in horoscopes, is not the position of any particular one of the asteroids; it is the position Phaethon would have occupied had it been entire. It is the echo of the origin point of the asteroids, and indeed of the solar system in its present form.

It is not easy at first to fit Phaethon into the astrological universe. There is a temptation to see it as 'Inner Planet 4', after Mercury, Venus and Mars; it is probably more useful to think of it as 'Outer Planet 0', preceding Jupiter, Saturn, and as many of the ones beyond as one cares to use.

Unlike the Sun, which is a positive and outgoing force astrologically, Phaethon must be seen now (that is, the shadow of itself) as a negative and collecting force. It can also be seen as a centre of disruption, a breaking-point, especially when in malefic configurations.

Each part of the human body, or bodily system, has a planet associated with it, and Phaethon can be fitted into the pattern without much difficulty. If Phaethon is to do with history and ancestry, then it must look after some part of the body which perpetuates one's ancestry, and which is not usually seen or referred to — Phaethon being now out of sight and out of mind. Everybody's ancestors have contributed to their own DNA; in fact, genes are immortal, passing as they do from generation to generation, adding to their history as they go along. Phaethon could well be connected with this. We tentatively suggest that Phaethon be connected with bone marrow and the cell formation processes therein. There are a number of symbolic connections, through the ideas of ancestry and the idea of the core, and it may be that Phaethon has a connection with diseases of this area, such as leukaemia.

The asteroids have long had an association with Pisces, possibly since they physically embody the processes of disintegration and dispersion associated with the final stage of the zodiac. It seems reasonable to associate Phaethon with the opposite sign of Virgo, and the idea has much to recommend it. The 'ecological' associations of Phaethon fit very well with the Virgoan emphasis on looking after oneself, eating carefully, and the proper use of available resources. It may not be too fanciful to assign to Phaethon the rulership of Virgo: if nothing else, it puts the two minor lights in a sign each side of the Sun's, and maintains a sort of balance, though this is fraught with problems of its own symbologically. Certainly Phaethon is in its detriment in its current state of being the asteroids, so assigning the detriment is no problem. This may be the first time that rulership has been determined from the detriment and not the other way round!

The same sort of thing can be done for the exaltation. We have no information about the exaltation of Phaethon, but there are a lot of stories about his fall, and we should look at them more closely. Ovid suggests in the *Metamorphoses* that the horses of the chariot of the Sun became frightened when they saw the claws of Scorpio in the heavens, and that it was here that Phaethon lost control. Whilst it is true that the *Metamorphoses* are essentially a work of literature, and that poetic invention is an essential part of the process, most myths cling tenaciously to the facts that they enshrine none the less, and Ovid is quite good at putting real data into verse, as some of his other works show. If the stars that he refers to in the constellation of the Scorpion are a reference to the zodiacal position of Phaethon at the time of its destruction, then the fall of Phaethon, in both literal and astrological senses, could be in early Scorpio; this would put the exaltation in Taurus. The fact that Luna's exaltation and fall are already here leads to some interesting speculation.*

It is hoped that the above will give a general framework of ideas associated with Phaethon, and the reasons why. It is important to establish the principle and the feel of its interpretation rather than to tie it to a list of specific items and to state that Phaethon is the

---

* See Appendix B, p. 169, for a discussion of this and a possible dating of the event.

ruler and significator of this or that. The next section develops these interpretative guidelines in more detail and gives some examples.

3.

PHAETHON INTERPRETED

Phaethon in Combination with Other Planets

To produce interpretative values for Phaethon in combination with the other planets, we have taken the planetary principles individually and combined them with the values assigned to Phaethon in the previous chapter. We feel that the concepts we arrived at from there are very indicative of Phaethon's peculiar retrospective viewpoint. These concepts may have a beneficial role to play in the life, or not, according to the nature of the aspect: Phaethon trine Saturn concerns the same sort of thing as Phaethon square Saturn, but is expressed in a different way.

Sun/Phaethon
The core of the spirit. This combination gives a sense of the individual identity as being a result of, and somehow linked to, the past. Each individual can be thought of as the apex of a pyramid: beneath him are his parents, and beneath them their parents, and so on. Whilst this is true for everybody, and if the process is taken far enough back it can be seen that almost everybody is descended from almost everybody who has ever lived, the idea is not often one which becomes a guiding principle in life. When Phaethon is in combination with the Sun, however, this idea is very much to the fore. There is a feeling not just that the person is different and individual, but *special,* a representative of the past with a vision for the future. In some instances, this vision for the future may indeed be a reconstruction of the past, an analogous process to the reformation of Phaethon from its scattered remains. There is a strong element of national or racial heritage in this combination: concentration of national identity is sought, rather than devolution, decentralization, or international

mingling. Phaethon is at 7 Taurus in the chart of Adolf Hitler, not far from his Sun in the first degree of that sign. Helena Blavatsky also has Phaethon conjunct her Sun, at 16 Leo. See also the charts of David Ben-Gurion and Kemal Atatürk on pages 47 and 49.

Moon/Phaethon

This combination is difficult to construct, for historical reasons. Luna, our present moon, may well have reached us in the period following the fragmentation of Phaethon, as a replacement. Phaethon was the original lesser light, a role which Luna now performs. To interpret the two planets together astrologically is in some respects like trying to imagine a planet in aspect to itself, and in other respects like an eclipse. Certainly the emphasis is on the intuitive and emotional processes in man; thoughts and feelings are doubly emphasized when both of the lesser lights are involved. Those thoughts and cares are retrospective here, as might be expected, with a sense of responsibility carried forward from the past and brought to bear upon the present. Things that were important in the past do not lose their relevance when more recent events come to the forefront of one's mind, but must still be given their due. There is a feeling that care is necessary to avoid catastrophe, as though Luna were a memorial, a reminder not to forget what has happened in the past. Care to maintain and keep good all that has been handed down from the past is indicated here, too; not just care of people and objects, but of standards and beliefs. A good word for this combination might be *conscientiousness*. Schumacher is a good example of this; see page 54.

It may be that thoughts and emotions from earlier eras are awakened and tapped by this combination, too: see the horoscope of Jung on page 52, and Schliemann below.

Mercury/Phaethon

Memory, the keyword, is easily arrived at, being the mental and intellectual processes associated with Mercury directed backwards in time. It would seem likely that this combination would produce an interest in history, and that a person with this combination natally might find his imagination stimulated by the idea of the distant past, or stories about it. Schliemann had the two bodies in a loose, but applying, opposition; the strengthening of the aspect by progression as he grew up indicated quite well his growing desire to dig up the cities he had read about in Homer.

Ancient languages may have some appeal: if a child shows a greater aptitude for Latin and Greek over French and German, then Phaethon may be in good aspect to the Mercury. It is unlikely that it will strengthen the intellectual capacity as such, but it will enable it to function with increased clarity and facility when directed towards the past rather than the future.

It is worth noting that Phaethon and Mercury do not produce deep and unconscious memory, such as race memory: the trans-Saturnian planets are a much surer indication of that. Mercury's concern is much lighter, primarily concerned with written history, and the languages used to express it.

Venus/Phaethon

As with the Moon, this combination is difficult for historical reasons. Venus may have been the intruder in our planetary system; whatever the reality, it seems to be no friend of Phaethon. Phaethon's rulership and detriment are Venus' fall and exaltation, though Venus' rulership is Phaethon's exaltation, so there is some affinity to be had at least. The historical relationship between Venus, Luna and Phaethon may help to explain their astrological relationship, but it is not fully established yet.

On one level Venus and Phaethon may point to a perception of the real values of things. This might be from the qualitative and quantitative facilities of Venus combined with the 'original and primary' flavour of Phaethon. It would seem likely that ecological interests would be represented by this combination.

On a lighter level (taking Venus as an air sign ruler rather than an earth sign ruler) a longing for the 'good old days', nostalgia for a bygone age, comes to mind. At the moment (1988) there is a considerable vogue for the restoration of Georgian and Victorian houses, and a general dislike of modernism. Some people affect the manners and dress of a pre-electronic age, as well. This seems indicative of Venus and Phaethon: the electronic age can be symbolized by Uranus, first planet beyond the traditional seven-planet system, so those fleeing from it will look in the direction of Phaethon, first planet from *before* the traditional system. The fact that Venus is part of this arrangement shows that only the enjoyable parts, the fashion and furnishings, of the bygone age, are to be adopted; the social ills, and the poor state of health that often accompanied them, are not.

Mars/Phaethon

In this combination Mars may be returning to its own Sun. Even if this is not so, any interpretation must involve the idea of the minor malefic exerting its power inwards and backwards.

Repression, depression, and guilt are the keywords of Mars and Phaethon. Mars is probably less malefic when orbiting its own Sun, but from a modern geocentric viewpoint the red planet is reminded of its loss when in aspect to the ghost of its parent. Emotions and passions are intensified, as is always the case when Mars aspects a light, and they are expressed only with difficulty, since the light to express them by is no longer there. This can lead to an obsessive state. Having the two bodies in good aspect may indicate that these feelings of repression and guilt do find some sort of expression: see the chart of the Marquis de Sade on page 52. In addition, this combination may contain the guilt of previous generations for their actions, especially if the breaking of Phaethon was not a natural occurrence.

Superstition and ritual may also be connected to this combination, in the sense that they are survivals from earlier practices. They are always *active* processes designed to bring alignment with higher forces, and are therefore connected to Mars; Phaethon gives them the retrospective direction necessary to prevent them developing or changing.

Jupiter/Phaethon

Jupiter is much happier with Phaethon, as all the outer planets are, since it may well have orbited both the Sun and Phaethon together, and the disruption to its orbit was much less than that undergone by Mars.

In interpreting the combination of Jupiter and Phaethon, some sort of single idea is needed which contains both the expansion and the progressiveness of Jupiter with the reflective and backward-looking qualities of Phaethon. The result will be all-embracing, if nothing else. Jupiter will take in all the ideas and events of the past which are offered to it by Phaethon, and use every one of them in constructing something enormous which will expand onwards into the future. The combination of these two planets, at least in their symbolic values, will give a result which stands outside of time: Jupiter will provide the future, and Phaethon the past.

There is an important process in the world of ideas which is provided by these two bodies in combination: that of implication.

Taking truths established in the past and moving them forward via imaginative reasoning is to infer from things implied; seeing the implication is Jupiter from Phaethon. Implication can have two meanings here, and the planets describe them both: firstly, implication meaning 'the likely future effect', and secondly implication meaning the weaving in of extra levels of meaning and relevance to the proposition currently under consideration.

A single example of a notable person who was born with Jupiter closely conjunct Phaethon will make the whole process clear. Einstein had Phaethon a degree before Jupiter, at 26 Aquarius.

Saturn/Phaethon

Saturn is not as malefic as Mars in combination with Phaethon, and the reason for this is the same as it was for Jupiter: it is an outer planet, and so did not lose the centre of its existence when Phaethon broke up. It maintains its general weight and pressure, though, bringing difficulty of movement to whatever it aspects. It is still lord of time, though, and also of structure: when added to Phaethon, one's ties to the past are made obvious and real.

A good word for Saturn and Phaethon is 'obligation'. The word itself has 'ties' in it (the -lig- part), and obligation is often something which has ties with the past and which hinders progress in the present. The links with the past in this instance are heavy ones, as befits the nature of Saturn; those of the Moon with Phaethon are similar, but are undertaken with a lighter heart, somehow. Both of them are serious in flavour, though, unlike Phaethon with Mercury, or Venus.

It appears to be impossible to take the idea of 'the structure of the past' too literally for this combination. Gregor Mendel, the monk whose work led to our understanding of heredity and genetics, had Saturn and Phaethon at square. They were in fixed signs: Phaethon in Leo, Saturn in Taurus. Is this not a reasonable symbolic reference to the structure of our genetic past being made definite? As an interesting parallel, Francis Crick, whose work with James Watson led to the discovery of DNA and its now-famous double-helix structure, had Phaethon and Saturn in sextile. See page 50.

Uranus/Phaethon

There is a strong element of discovery in this pairing, which is not surprising. There is also, often, the manic belief in one's own viewpoint: this is the man who proudly exclaims that the rest of the regiment is out of step with him. This can lead to a certain unpopularity or lack of preferment, though history and posterity may well reappraise the situation with hindsight, which is of course itself Phaethon's own contribution to perception. Kipling had Phaethon in the second degree of Pisces, trine to Uranus at the same degree of Cancer; his views and his writings on India were often unpopular and seen as critical of the British presence there. Helena Blavatsky, author of *The Secret Doctrine* and founder of Theosophy, had Phaethon and Uranus at opposition. So did Einstein — the difference in the planetary pattern is that Blavatsky's horoscope has Jupiter with Uranus, both opposite Phaethon (and the Sun), while Einstein has Phaethon with Jupiter, both opposite Uranus. Both came to produce enormous, all-encompassing versions of reality which were highly unorthodox in their time. See also Nietzsche, page 54.

Phaethon and Uranus seem to be in aspect in the horoscopes of quite a number of scientists. Wilkins, who worked with Watson on DNA, has Phaethon and Uranus in trine.

Neptune/Phaethon
'The inheritance of dreams'. This combination links the history of the original planetary system to the subconscious memory of it. Individuals with these two bodies in aspect might reasonably be expected to be able to use their dreams as a communicative tool. Edgar Cayce, the American psychic, had Neptune and Phaethon in trine (Phaethon 29 Leo), which comes as no surprise.

Shamanistic practices, and any others in which the practitioner deliberately forsakes his conscious mind for the subconscious, are connected to Phaethon and Neptune. In most cases there is a strong emphasis on contacting the spirits of ancestors or race leaders, whose knowledge is clear and whole; all of this seems strongly indicative of Phaethon.

In some respects Neptune and Phaethon together is a contradiction in terms. Phaethon was once concentrated, but is now dispersed; to follow its call backwards in time is to reconcentrate it. To do this with Neptune, the symbol of diffusion and dispersal, is difficult to visualize. Neptune is representative, as the ruler of Pisces, of Phaethon in its fragmentary state; Phaethon, of course, represents itself whole. Thus the two bodies must always act as though in opposition. The core of reality lies at the heart of each fantasy or dream, and any given moment can have many different origins and outcomes. Phaethon and Neptune represent the interchange between the formed and the unformed.

Pluto/Phaethon

If Pluto is, as has often been suggested, the principle of organization, then Pluto and Phaethon must be reorganization after disintegration. Individuals with this configuration may well have the capacity to restructure things which have been lost, scattered, or forgotten in the past, and to do it entirely through their own efforts. Jung (page 52) has Phaethon in the middle of a Moon-Pluto conjunction, as good a metaphor as one could devise for the restoration of the universal psyche and its links with the past.

Pluto and Phaethon must also have some connection with large-scale disasters; if their effects are irreversible, yet their occurrence could have been prevented, then the symbolic values of both bodies are well represented. They were in close opposition on the occasion of the Chicago air disaster of 25 May 1979.

Phaethon in the Signs of the Zodiac

To produce interpretative values for Phaethon in the various signs of the zodiac, we have been guided by three things. Firstly, by the principles of the signs themselves; secondly, by the 'house sequence' of the signs, assuming Phaethon to be in its own first house in Virgo; and thirdly, by world events in the years when Phaethon stayed most of the year in a particular sign. Although Phaethon moves on average just over 79 degrees annually,* there are some years in which it seems to spend a disproportionate amount of time in one sign. A good example is 1964, where it started the year at 21 Scorpio and finished it a mere 46 degrees away at 8 Capricorn, having spent the period from late January until mid-November in Sagittarius. This is almost the sort of orbital rate one might expect to see from Jupiter! These 'long-stay' years seem to recur at regular intervals, and we feel sure that they will make their presence felt in mundane astrological work.

Phaethon Aries
This position is very powerful, as seems to be the case when Phaethon

* mean annual rate 79.47°; orbital period 4.53 years.

is linked with Mars in any way. Aries gives new beginnings and the initial impetus to things, while from Phaethon's point of view it is in its own eighth house. There is an emphasis here on breaking with the past, and on starting again. It may be that under this placement events develop so quickly that they exceed their natural pace, and destroy themselves — a sort of 'boom and bust' phase, after which reconstruction must take place. Certainly the eighth house connection suggests that events signified by Phaethon in Aries are of an irreversible nature: that is, they are of the greatest importance, and once the event is past, life is not the same again, nor can it be made so. Historically, the years when Phaethon has been in Aries have been the ones when world finances have collapsed after a preceding boom period: the Wall Street Crash of November 1929 is a good example. 1988-9 looks similar, though the effect of the few months spent in early Taurus may change the situation.

Phaethon Taurus

This is Phaethon's exaltation, and its own ninth house. The emphasis of this position is rest and recovery from past exertions and sufferance. Inspiration and energy to face the challenges of the future are found here, and a chance to regain self-composure before moving on. Earth sciences should have their home here; anything which uses the Earth as a source of comfort and nourishment, but with a care for the earth's own history and responsibilities, is showing the qualities of Phaethon in Taurus. Ernst Schumacher, whose 'one small planet' ideas first popularized the notion that the capacity of the planet to give profit and suffer abuse was not infinite, has Phaethon at 3 Taurus, which may be its exaltation degree.

Mental rest and recovery must be included here too. Jung had Phaethon in Taurus, conjunct his Moon at 19°. There is much to suggest that this great pioneer was not working so much with the lunar side of human nature as with the fragments of Phaethon's side, in its role as the original lesser light.

Phaethon Gemini

This placing must be similar to the Phaethon-Mercury combination in many ways; it concerns itself with the knowledge of the past, and with previous communication of all sorts. History and mythology are part of the expression of Phaethon in Gemini: Schliemann, whose Moon and Phaethon we have noted before, had the pair of them here. It was, after all, the appeal of precisely those two qualities, history and mythology, and the belief that in some cases the two were identical, which spurred him to his eventual discoveries at Troy and Mycenae.

Since this is also Phaethon's tenth house, there must be some sort of lasting reputation to be made here, a chance to become a household name. Queen Elizabeth II has Phaethon in Gemini, and so does Margaret Thatcher; Phaethon was a long time in Gemini in 1925-6. Both women are assured of their place in the history books.

Phaethon Cancer

Here the old lesser light is in the sign of the new one. This placement brings tears and upset, with sensitivity heightened to the point of neurosis. Feelings are on the surface here, and a person with this placement natally will be very easily wounded emotionally.

In this position Phaethon can look backwards and forwards at the same time with equal sorrow, remembering the suffering and anguish of the past as it anticipates similar events recurring. It was in this sign from the autumn of 1939 to the summer of 1940, the opening months of the Second World War, and it was slow in Cancer again at the end of 1957 and the beginning of 1958, a period which in Britain at least saw, amongst other things, the development of the hydrogen bomb, the opening of the first nuclear power stations, the foundation of CND, and the first Aldermaston marches.

Being Phaethon's eleventh house brings hope amidst the tears, because it looks to this sign to provide it with ambition for the future. In Phaethon's absence, Luna has to work on its behalf, to be the visible representative of the ghost of its predecessor, and to be the light for those times when the Sun is not visible.

Phaethon Leo

This placing finds Phaethon in the house of its former partner-and-enemy, as its binary star partner must have been. It is also its own twelfth house, of course. There will be a strong element of jealousy apparent with Phaethon here, and anyone with a strong aspect pattern to Phaethon in this position may find themselves involved in a love-hate relationship, where to part is impossible but to continue leads to heartbreak. This is exactly what happens astronomically to binary star systems: the two elements are attracted to each other, but their energies are set against one another, and usually the larger one destroys the smaller one in the long term.

The idea of jealousy suggests fighting back rather than merely suffering the loss, and Phaethon in Leo has a competitive nature to it, an enjoyment of the struggle for ultimate mastery. Phaethon is at 3 Leo in the horoscope of Nietzsche.

Phaethon Virgo

This is Phaethon's rulership, and the sign in which it has the best opportunity to show its true nature. All of the qualities cited earlier in 'Phaethon in Astrology' apply to the planet in this sign. It is careful, conservative, and reflective by nature. It is most clearly shown in the roots of things (Virgo is an earth sign) and at a very small and elementary level (Virgo has an affinity with the digestion processes; its energy works at the level of enzymes and proteins). There is an interesting polarity problem in that Virgo breaks things down to process them, and Phaethon would, one assumes, rather not; it might be argued, however, that Virgo breaks down, as all mutable signs do, in order to reform in a better state, and that Phaethon's story has a similar message.

A key phrase for Phaethon in Virgo is 'the memory of loss'. This seems to catch some of the essence of the body's astrological meaning. We know of an individual with Phaethon in Virgo who lost both his mother and his twin at birth; obviously this had a lasting effect on the development of his character. It is also worth noting, in the light of our key phrase, the fact that Phaethon was in Virgo for relatively long periods in 1918 and 1945, the closing stages of the great wars in Europe this century. It is on such occasions that nations have time to assess the cost of their endeavours.

Phaethon Libra

This is the environment of the past, made whole again. Peoples and territories which were formerly collected in one place are re-assembled from their fragments, and the arguments, wars, and divisions which broke them up in times past are settled. Libra's connection with peace and harmony has a definite role to play in this process. Through Libra and the action of its ruling planet Venus, the past can be reconciled with the present and have a place to live and grow towards the future.

People with Phaethon in this position natally are pacifiers and unifiers; their business is to find the common ground between two sides, and to bring them together again.

David Ben-Gurion has both the Sun and Phaethon in Libra. He represented the reunification of his people, and to a great extent the re-creation of their homeland in the modern state of Israel.

Phaethon Scorpio

Phaethon is powerful and destructive again here, in a sign ruled by its erstwhile satellite, Mars. It is also in its fall, the sign it was in at the time of its destruction. It is not usual to find destruction as a distinctive feature of a planet's action in its third house, as here, but there is usually movement connected with third house matters, and Phaethon in Scorpio shows great movements in an historical sense.

Great changes of the state of things, when the old order is swept away for ever, are shown by Phaethon in Scorpio. They are symbolized not only by Phaethon itself being swept away, and the massive reorganization of the planetary system which followed, but also by the forceful and searching nature of the sign itself. Somehow Scorpionic changes seem to be more thorough, to delve deeper, and to have more lasting consequences than others; the very core of things seems to be uprooted in such events.

Phaethon occupied Scorpio for much of 1914, up to and including the outbreak of the Great War; the structure of English society was never the same again, and the old order could not be re-established.

Phaethon Sagittarius

Here is the New Order, established after the cataclysms of the preceding sign. Phaethon is in its own fourth house, re-establishing itself under the care of Jupiter (as a belt of asteroids?). Phaethon has been spending long periods in Sagittarius at nine-year intervals recently: 1946, 1955, 1964. The sequence can be seen forming in 1928 and 1937, but Capricorn is involved as well, and the effect is not so marked; Scorpio is involved in a similar way as the sequence fades out in 1973 and 1982. In 1946 the world was struggling to rebuild itself after the Second World War, while in 1955 it was trying to shake off its immediately 'post-War' image of itself, and look forwards. The events of 1964 are very marked indeed — a Labour government in Britain for the first time in 13 years, the arrival of pop culture, the complete dominance of popular music by the Beatles, and much else besides. The 'Sixties as an era came into its own with Phaethon in Sagittarius in 1964.

Phaethon Capricorn

This is Phaethon's fifth house, and Saturn's first; Phaethon is heavy and sombre here. All of the earth signs seem somehow to be connected with the physical experience of loss when Phaethon is in them, and here in the cardinal earth sign is found the simple physical reality of the event, the sudden hole at one's side caused by loss itself. Virgo is the memory of loss, and hopeful Taurus is recovery from loss. The rulers of the signs give clues to the working of the symbolism: Mercury and Saturn are not bright enough to rekindle the flame of hope after loss, but Venus, especially as a morning star, when it is given the title Lucifer ('light-bringer') to tell of the imminent new dawn, has light enough for the job. Thus Venus' earth sign gives hope of recovery, but the other two signs are unable to lift their spirits. Mercury's earth sign is concerned with the mental aspect of loss, while Saturn's deals with the actual fact.

Phaethon stayed for long periods in Capricorn in 1901, 1910, and 1919. Queen Victoria died in 1901, and the nation went into mourning, unable to cope with the absence of a ruler who had seemed almost eternal. Her son, Edward VII, died in 1910. In 1919 the first Armistice Day services were held, and almost every village has a war memorial with 1919 inscribed upon it as the date of its building. Memorial stones are, of course, a very Saturnine thing. Reference has been made to the end of the Great War before, under Virgo: but the fact remains that in 1918 and 1919 Phaethon spent much of its time in those two signs, while covering the trine between them with almost indecent haste.

Phaethon Aquarius

This is the group from the past, reformed in the present. Aquarius has always been associated with large groups of people, and especially large groups whose motivation is a shared political belief. Phaethon's action is to take a common origin point, or common heritage, and make that the basis for a modern movement, so that the past can live again, reformed and reunited.

Aquarius is also the sixth house of Phaethon, and there are therefore connections to be made to matters of health and public service, but also to the public guardians of a sovereign state, and indeed to its armed forces. Much has already been said about the links between Pluto and the Nazi party, and rightly; but Phaethon deserves consideration in this respect as well. Phaethon spent much of 1933 in Aquarius, when Hitler, as Chancellor, advanced his own position, and that of the party, very quickly and very successfully. Although Phaethon was still in Capricorn when Hitler officially became Chancellor, its position on that day makes no fewer than six major aspects to planets and angles in the chart for that event, and it was just about to enter Aquarius when the Reichstag fire occurred. The unique flavour of National Socialism — a political movement formed on modern and slightly mechanistic lines, but fuelled by the appeal of an ancient race mythology — is precisely expressive of Phaethon in Aquarius.

Those born with Phaethon in this sign and strongly aspected should be firm fighters for causes with their roots in the past, or for causes which oppose the abuse of our common heritage. Man and the planet have common roots: 'Aquarian-ness' and 'sixth-house-ness' seek a healthy future for them both together.

Phaethon Pisces

This is the sign of Phaethon's detriment. It reflects Phaethon's loss of power, and its fragmentation. Phaethon shows itself here in breakups and breakdowns of all kinds, where links with the past are deliberately cut or allowed to fade and die. There is failure from lack of concentration, from knowledge and resource being spread so widely as to be ineffective and wasted.

Historically, Phaethon has been prominent in Pisces in 1965 and 1974. In each case there are plentiful examples of major powers losing control of distant territories. Rhodesia declared itself independent of Britain in 1965, while in 1974 American interests suffered serious reverses in South-East Asia.

Phaethon in the Houses of the Horoscope

In devising these interpretations we have attempted to blend the usually accepted principles and symbolic values of the houses with those we have posited for Phaethon. We have also allowed ourselves to take inspiration from the house position of Phaethon in the horoscopes of prominent people.

The sequence of positive/negative polarity seems to be at least as marked in the houses as it is in the signs. Odd-numbered signs/houses seem to show Phaethon dispersing energy, a parallel process to its own fragmentation. Even-numbered signs/houses present the opposite side: Phaethon is a concentrated and collected presence from the past. The idea of *loss*, however, is present irrespective of polarity.

Phaethon First House

The embodiment of the past. Those who have Phaethon rising are in some way literal embodiments of their own history. This can be their genetic history, their family history, their racial history, or any other kind of history, but somehow their physical frame ties them to their ancestry. A most striking example of this is Simon Wiesenthal (page 46). Also worthy of note is Pierre Teilhard de Chardin, the palaeontologist and Christian philosopher, whose horoscope has Phaethon in the first house too.

Inherited diseases are probably connected with Phaethon here. It would not be too fanciful to suggest, given that the lesser light is associated with the psychological and emotional side of being, that inherited psychoses are also indicated by Phaethon in the first house.

Phaethon Second House

Rags to riches, and back again. Obviously a second house Phaethon is going to make its presence felt in the financial affairs of the native, but we feel that it can indicate both a dispersion and scattering of money, and the collecting and amassing of wealth by using various links with the past. There is often the idea of the 'family fortune' somewhere; we have numerous examples where the native lives entirely on inherited wealth, precluding the need to develop a career or any particular skills to provide him with money (Phaethon in the second affecting the other two houses of its triplicity, the tenth and sixth).

On a larger scale, Phaethon in the second suggests the preservation of the old financial system, and the use of the resource held within it to finance the future. President Roosevelt had Phaethon in this position; his famous New Deal saved the old financial system from complete dissolution, and gave it a new face for the future.

Phaethon Third House

This position seems to give an awareness of the passing of time, and of opportunities and ideas whose time has come. Edison had a third house Phaethon.

The loss of siblings, close relatives, and neighbours seems a reasonable interpretation for Phaethon here, especially if their influence was such that their relationship with the native and the circumstances of their loss shape his later behaviour patterns.

The third house's special affinity with all forms of communication suggests that ideas from the past can be reborn and reformed through this house, and communicated from here to new generations. Mussolini had Phaethon here; his idea was to rebuild the Roman Empire in the twentieth century, and he was an accomplished orator.

Phaethon Fourth House
Phaethon here deals with the destruction, dispersal, and eventual re-establishment of the family home and its cultural roots. We know of an individual who has been exiled twice, from different places, during his life, and who has Phaethon in the fourth house in Aries. The Aries gives him the impetus to start again, but it is the fourth house position which shows how his nationality and his past have conspired to deny him a home in the present.

Mao Tse-tung had Phaethon in Taurus in the fourth. His feeling for the land of China, and his re-establishment of the nation as an agrarian People's Republic are simple but precise parallels to the position of Phaethon in his horoscope.

Phaethon Fifth House
The representation of the past. Where the first house Phaethon stressed the genetic links with one's ancestors, here the links are more in representative actions. Queen Elizabeth II has Phaethon in the fifth house, and can be seen continuing in the role of monarch, taking part in the ongoing tradition.

There is a certain theatricality to this position, and a sense of showing the symbols of the past to future generations. Both Joseph Smith and Brigham Young, the key figures in the foundation of the Mormon Church, had Phaethon in their fifth houses. In Smith's case, Phaethon is also conjunct Jupiter, to provide him with his all-encompassing vision and inspiration.

Phaethon Sixth House
Phaethon's influence on the systems of the body must be uppermost here. Inherited diseases are likely to be indicated by this position, as are any problems of a genetic nature, or such as are caused by

genetic mutations. Phaethon seems to have a special affinity with the bone marrow, and the formation of cells; the condition of, or imbalance of, the various cells in the bloodstream is also within its influence.

Away from the body, Phaethon's influence in the sixth house may include responsibilities reaching back into the past, or to previous generations, such as the care of older people. It will also indicate any interest in the arts and crafts of previous eras. To actually learn and practice a skill from the past, simply to keep it alive, would be very indicative of Phaethon here.

Phaethon Seventh House

Phaethon returns to the individual and his view of himself here, as befits its position at the other end of the physical angle. Rather than embodying the past, as was the case with Phaethon rising, the individual sees himself as being in conflict with the past. The enemy is seen as being any traditional practice or established order which is upheld simply because 'things have always been done this way', and the individual sees it as his duty to challenge this. In his view the past must make way for the present and the future, and the good things which are held in check by restrictive practices must be liberated so that they can play a fuller role in times to come.

Hitler had Phaethon in the seventh house, suggesting that he wanted to break Germany away from what he saw as the restrictive obligations imposed by the Treaty of Versailles and to make the best use of the talents of the German people.

Havelock Ellis, the American sexologist, had Phaethon in the seventh too. Here the desire is to lift the restriction and taboo of previous generations and promote a more open discourse about sexuality.

Phaethon Eighth House

People with Phaethon in the eighth house are those who delve into the past and examine the origins and consequences of things. They are adept at gathering together separated elements and concentrating them into a powerful and effective whole. Often there is a feeling of 'destiny' about them, as though the native feels in some way destined to make his mark on history through the intensity of his beliefs and the strength of his character.

General de Gaulle had Phaethon in the eighth house in Libra.

Although not at first recognized as the leader of the French after the German invasion of 1940, his belief in himself and his ability to reassemble the separated elements of France led to his eventual success.

Margaret Thatcher has Phaethon on the eighth house cusp, and the astrologer Dane Rudyhar had it in the eighth too; very different people, but both with a strong belief in their own viewpoint and methods.

Phaethon Ninth House

The inspiration of the past. Prominent people with Phaethon in the ninth house seem to personify reconstruction and realignment. Traditional values are given a new twist and made to seem fresh and vital again, and there is hope for a stronger and better future. The idea of a 'traditional future' may seem like a contradiction in terms, but Phaethon does not take its hope from technological development; it takes solace from remembering the past.

Ronald Reagan, Mikhail Gorbachev and David Ben-Gurion all have Phaethon in the ninth house of their horoscopes. All, to some extent, based their political appeal on the idea of 'together, like we used to be' — an idea in which Phaethon's influence is plain to see.

On a non-political level, H. Rider Haggard, the novelist whose tales of mysterious cities and ageless princesses were once so popular, also had Phaethon in this position. The key elements of his stories — their location in hidden lands known only in myth, the idea of myth becoming actual fact, the strong mystical element — are all indicated by Phaethon in the ninth house.

Phaethon Tenth House

The tenth house of a horoscope represents, in many ways, the way the world remembers an individual. Reputation is portrayed here, which is often very different from the individual's real essence; none the less, it is for his reputation that he is remembered.

Phaethon's role as the lesser light has been taken by Luna, whose presence obscures Phaethon's memory; the body itself has been fragmented and strewn around the solar system. We must dig deep into mythology to find Phaethon's history; most of the time it is far from our mind. The reorganization of the solar system is what we are presented with, and what we remember.

Those with Phaethon prominent in the tenth house seem to have the capacity to reorganize history, to present a different version of it which then becomes the accepted norm. Such people are always seen as founding fathers, paternalistic figures by whom the order of things was first determined. Although it is easy to see that there must have been an accepted viewpoint before these people, it is almost never mentioned: their status as creative demiurges is absolute. Henry VIII is a good example, with his reorganization of the Church in England. More recent examples include Sir Fred Hoyle, the astronomer whose theories on the origin of the universe are now the standard, and Adolf Eichmann.

Phaethon Eleventh House

The idea of dispersal seems to be the facet of Phaethon that is most prominent in this house. Ideas are taken from the past and used for the future, but somehow run away from their owners. It is the companions and colleagues of those with Phaethon in the eleventh who seem to get the benefit, and not the natives themselves. Phaethon here seems to be surrounded by opportunistic friends, which may make some comment in itself about its destruction; certainly Phaethon is let down by them, and has to struggle on as best it can. A good adjective for the feeling this produces is perhaps 'rueful'; misplaced trust teaches hard lessons.

The great inventor Tesla had Phaethon here, perhaps as expected, given the correlation between the eleventh house and the eleventh sign. He was not a commercial success, though; one of his best ideas, alternating current, was adopted by Edison, who went on to make a great deal of money from it, despite the fact that he had previously been a loud critic of Tesla's theories.

'Buzz' Aldrin, the second man on the Moon in modern times, has Phaethon in the eleventh house, square to his rising Pluto among other things. He was to have been the first man on the Moon, but somehow Neil Armstrong acquired the honour, and went into the history books. (The astrology of all of the Apollo astronauts and the times of their landings, with special reference to the positions of Luna, Lilith and Phaethon, merits careful study and is to be recommended. See Appendix 2 for the relationship of these three bodies.)

Phaethon Twelfth House

The obvious inference to be drawn here is that the individual puts the past behind him. Whilst this is quite true, there are additional subtleties to be observed: the actual relationship with the past is a love-hate one, so that although the individual is constantly trying to free himself from the constraints of the past, he is very attached to it at the same time, almost to the point of obsession.

Anybody who is at such pains to put his past behind him will be seen as a rebel by the rest of society, which is, on the whole, glad of the support that its past affords it.

The twelfth house is often associated with the care of the sick, chronic diseases, and with institutions: it is interesting that Pasteur had a twelfth house Phaethon. It is possible to see him, through the Phaethon position, as attempting to defeat Man's eternal secret enemies from the past: the viruses which follow us down the generations match us new genus for antibody, antagonists for evermore and yet dependent on each other for development and survival.

W. B. Yeats had Phaethon very late in his twelfth house, practically on his ascendant. Assessing him as a twelfth-house rebel makes sense, but some may prefer to see him as having Phaethon rising. Like Weisenthal, our example for the first house, his Phaethon stands astride the start and finish line of the circle of houses. Obviously the past is of great importance to these people, but society finds their precise use of it difficult to gauge.

4.

PHAETHON EXAMPLES

Simon Weisenthal

Simon Weisenthal has devoted the last 40 years to tracking down German war criminals. Such dedication is not immediately apparent from his horoscope. The Sun is angular, and Saturn is loosely square to it; a steady and determined approach to work could be suggested, but hardly enough to justify his single-minded pursuit, even if the Sun's fourth house placing is taken to indicate a concern for one's family and its origins.

Adding Phaethon to the horoscope makes a big difference. Phaethon is conjunct the ascendant: this man feels that he is the history of his people incarnate. The life of every one of his ancestors is contained in him, and he is very conscious of it; it is not simply a question of feeling for the past, but of actually being the past. He can never be parted from the sensation of having lost his forebears, because he feels that he is a literal continuance of them.

Phaethon is in opposition to Saturn here: he has a duty to his past. Ordinarily it would be enough to be conscious of the debt one owes to one's ancestors, but with Saturn involved there is an obligation to repay that debt. The obligation is no trivial matter — Saturn is a heavy planet — and it will involve sacrifices and hardships, as the square to the Sun will testify.

The power of this horoscope, and of Weisenthal's character, lies in the angularity of the Sun-Saturn-Phaethon T-square: whichever corner of the triangle is considered first, there is something there to bind Weisenthal back to his past, and to the sufferings of his people.

David Ben-Gurion

This makes an interesting contrast to Weisenthal. Instead of Phaethon

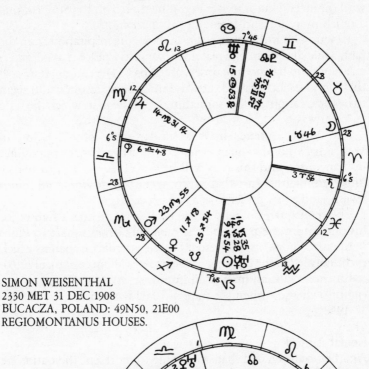

SIMON WEISENTHAL
2330 MET 31 DEC 1908
BUCACZA, POLAND: 49N50, 21E00
REGIOMONTANUS HOUSES.

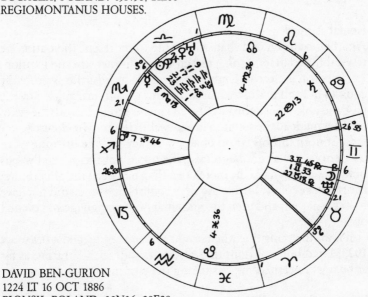

DAVID BEN-GURION
1224 LT 16 OCT 1886
PŁONSK, POLAND: 52N36, 20E29
REGIOMONTANUS HOUSES.

with the ascendant, it is here in conjunction with the Sun, making a much more forward-looking combination.

Phaethon's ninth house position shows the inspiration from the past, while the Libran sign placement suggests a peace-making effort, an attempt to balance the upsets of the past. The conjunction with the Sun gives a creative desire: the most likely result would seem to be the creation anew of something from the past, which the state of Israel was.

The Saturn square shows a similar sense of obligation to Weisenthal's, but the square is more constructive than the opposition in many cases, and this may well be one of them. It must not be forgotten in either case that Saturn is the significator of all things Jewish, and so its presence is to be expected.

The semisquare to Mars in the twelfth made him a fighter for his cause from the past, and the matching sesquisquare to Pluto in the sixth shows his involvement in the guerilla campaigns which eventually led to the uprising of the people (Moon conjunct Pluto).

For those with an interest in mundane matters, Phaethon was conjunct the Sun when the state of Israel was proclaimed on May 15 1948.

Kemal Atatürk
Another example of a man who came to represent the nation he re-established in something resembling its former size and location.

Phaethon is close to Uranus in the fifth. Whilst this is probably enough in itself to show a man with strong opinions, guided by the past and able to lead the way, the real interest must lie with the series of trines made to the stellium in the twelfth house. The closest aspect is to Chiron: did his vision of a new Turkey include learning from past errors and attempts to raise the quality of life generally? His new Turkey was markedly more open than had been the old Empire: the privileges of the aristocracy were abolished, women did not have to wear the veil, and even the grammar of the language was evened out.

It is worth noting that when he was elected President on 29 October 1923, Phaethon was transiting his north node in Sagittarius: as far as he was concerned, he was taking the past in the right direction.

Watson, Crick and Wilkins
James Watson, Francis Crick and Maurice Wilkins were the men

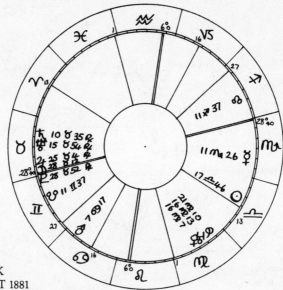

KEMAL ATATÜRK
1828 GMT 10 OCT 1881
SALONIKA: 40N38, 22E55
REGIOMONTANUS HOUSES.

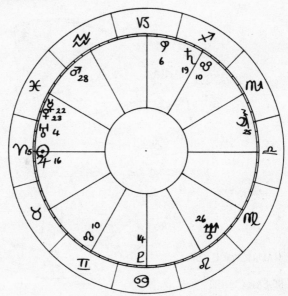

JAMES WATSON
6 APR 1928
SOLAR CHART.

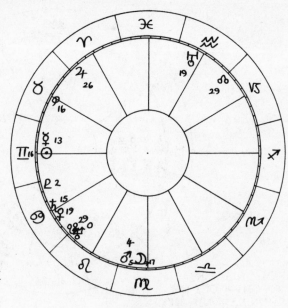

FRANCIS CRICK
8 JUN 1916
SOLAR CHART.

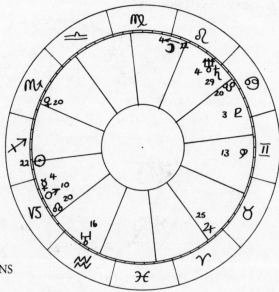

MAURICE WILKINS
15 DEC 1916
SOLAR CHART.

responsible for the discovery of DNA, the complicated substance whose famous double-helix structure enables coded genetic information to be stored within it.

The horoscopes as given are solar, with positions as at midnight. They are none the worse for that, of course: the outer planets are not sensitive to the hour of the day regarding their degree. In all three cases, it is interesting to note the patterns formed between Uranus, for discovery and inspiration (and science, too, possibly), Saturn, for structure and form, and Phaethon, for the genetic inheritance.

Watson's horoscope shows a loose T-square, with a stellium containing the Sun, Jupiter and Uranus at the focus; Phaethon and Pluto make the arms. The closest aspect from Phaethon, in the ninth, is the square to Uranus, emphasizing the idea of inspiration and discovery. From a solar point of view, Phaethon is the highest planet in the horoscope: it represents Watson's inspiration, aspirations and achievements.

In Crick's horoscope the aspect to Uranus is wider, but the closer aspect to Saturn from Phaethon compensates. Here the genetic inheritance of Phaethon is given real form and shape. Where Watson is weak, Crick is strong; the two men complement one another. In Crick's horoscope there is an interesting network of forming aspects between the Sun, Saturn, and Phaethon as they close on Uranus; all of them are in the fourth face of their signs, indicating the physical form of something universal. In addition, the Sun is on the Saturn/Phaethon midpoint, as good a pointer as one could wish for to Crick's most significant work.

Wilkins has Phaethon trine Uranus, in contrast to the squares of the other two; the harsh semisquare to Saturn provides a measure of balance. The astonishing thing about all three men is that Phaethon is in aspect to Saturn or Uranus, or both, in each case. The symbology of the discovery of the structure of DNA is quite openly on view.

A final point to ponder: it may be to do with the mutable nature of DNA, or to do with the now hidden nature of Phaethon, but in each case Phaethon is in a cadent solar house, as though hiding from the Sun.

The Marquis de Sade

A good example of a Phaethon-Mars contact. Although he gave his

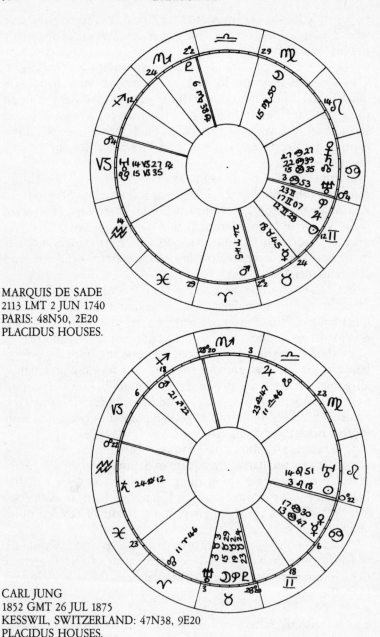

MARQUIS DE SADE
2113 LMT 2 JUN 1740
PARIS: 48N50, 2E20
PLACIDUS HOUSES.

CARL JUNG
1852 GMT 26 JUL 1875
KESSWIL, SWITZERLAND: 47N38, 9E20
PLACIDUS HOUSES.

name to sadism, de Sade was first and foremost a writer: he has the Sun, Jupiter and Phaethon in Gemini, which would predispose him towards the written word. It is Mars, though, and not Mercury, which is at aspect with Phaethon: the thing which bears his name down the centuries is not a book but a physical practice.

Jung

Jung's horoscope has been widely discussed in recent years. The conjunction of the Moon and Pluto is a beautiful representation of the collective unconscious written in planetary language, and Phaethon's position neatly fitting between the two serves only to underline their importance and relevance. If Phaethon really does represent the experiences of previous generations handed on, then for it not to aspect the Moon-Pluto conjunction would have been disturbing; to find it where it is in such a significant chart is very satisfying.

Other Nativities referred to with reference to Phaethon

Name	Date	Time	Place	Phaethon
Aldrin	20 Jan 1930	14.17 EST	40N48, 74W12	17 Ar.
Blavatsky	12 Aug 1831	02.17 LT	48N29, 35E00	16 Le.
Cayce	18 Mar 1877	15.30 LT	36N52, 87W26	29 Le.
De Gaulle	22 Nov 1890	11.54 LT	50N33, 3E05	5 Li.
Edison	11 Feb 1847	23.40 LT	40N00, 82W00	26 Cp.
Eichmann	19 Mar 1906	09.00 MET	51N10, 7E04	18 Aq.
Einstein	14 Mar 1879	11.30 LT	48N23, 10E00	26 Aq.
Elizabeth II	21 Apr 1926	01.40 GMT	51N30, 0W07	23 Ge.
Gorbachev	2 Mar 1931	22.08 LT	45N02, 42E00	20 Cp.
Haggard	22 Jun 1856	05.04 GMT	52N40, 0E50	29 Aq.
Havelock Ellis	2 Feb 1859	08.30 GMT	51N22, 0W00	22 Vi.
Henry VIII	28 Jun 1491	09.45 LT	51N28, 0W00	17 Cn.
Hitler	20 Apr 1889	18.14 LT	48N15, 13E00	7 Ta.
Kipling	30 Dec 1865	16.53 LT	18N55, 72E50	1 Ps.
Mao Tse-tung	26 Dec 1893	06.16 LT	28N00, 111E0	3 Ta.
Mendel	22 Jul 1822	?	54N01, 11E39	11 Le.
Mussolini	29 Jul 1883	14.00 LT	44N13, 12E02	3 Aq.
Nietzsche	15 Oct 1844	10.07 LT	51N17, 12E10	3 Le.
Reagan	6 Feb 1911	13.53 CST	42N20, 97W50	27 Aq.
Roosevelt, F.D.	30 Jan 1882	20.41 EST	41N06, 74W00	26 Li.
Rudhyar	23 Mar 1895	00.42 LT	48N40, 2E20	17 Le.
Schliemann	6 Jan 1822	?	54N01, 11E40	25 Ge.
Schumacher	15 Aug 1911	?	50N43, 7E06	3 Ta.
Smith, Joseph	23 Dec 1805	18.00 LT	43N47, 72W26	22 Sg.
Teilhard de Ch.	1 May 1881	07.00 LT	45N47, 3E05	16 Cn.
Tesla	10 Jul 1856	00.00 LT	41N29, 24E46	28 Aq.
Thatcher	13 Oct 1925	08.57 GMT	52N55, 0W38	17 Ge.
Yeats	13 Jun 1865	23.56 GMT	53N24, 6W20	23 Aq.
Young, Brigham	1 Jun 1801	11.00 LT	42N47, 72W52	28 Cp.

PART TWO: DARK SUNS

5.

DEFINITION AND HISTORY

Raymond Henry

The planets travel round the Sun in ellipses, not circles, and have done since the day we lost our companion star, Phaethon. An ellipse is the locus, or track, of a third point, such as a planet, around two focal points; a circle, by comparison, has only one. This single fact, that an ellipse has two focal points, has enormous implications for astrology, and yet it has been all but ignored since the time of the ancient Egyptians.

In the diagram below there is an ellipse with two focal points X and Y. The centre, halfway between them, is not important to our reckoning. The two lines XP and YP vary inversely: as one lengthens, the other shortens, and their combined length is always equal to

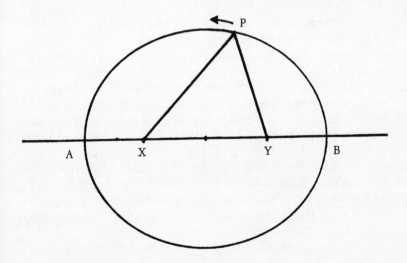

the 'major axis' of the ellipse, AB. If we take a mean of the distances AX and XB, then we have the 'mean radius' of the ellipse, the distance that would be the radius if the ellipse were a circle. If we subtract the mean radius from XB, and then divide the result by the mean radius, we have the 'eccentricity' of the ellipse. If we multiply twice the mean radius by the eccentricity, then we have the distance between the two foci, the distance XY in the diagram — and it is this that interests us.

Each planet in the solar system orbits in an ellipse which is different from the others: not only does the size, and therefore the distance from the Sun, vary, but the eccentricity varies as well. Venus's orbit, for example, is quite close to circular, and so is that of Neptune; Mercury, on the other hand, describes a very eccentric path, and so does Pluto, whose orbit is so eccentric that it is closer to the Sun than Neptune at certain times.

Each of the planets' orbits has a common focus: the Sun. In other words, if the two focal points for each of the orbital ellipses are calculated, one of them will always be the position of the Sun. This is not surprising — it is how the solar system functions, after all. But what about the other focal point, in each case? There is no star at these points — they are simply locations in space. Yet they are more than that, because a planet orbits each one as surely as it orbits the Sun. If, in making astrological sense of a planet's position in the zodiac, we take into account the star it orbits, then we must also take into account the second point it orbits as well, since its way through the heavens is determined by two points, not one. From each planet's point of view, there are two 'Suns'; the first is the star it shares with all the other planets, and the second is a point in space all its own.

It is a relatively straightforward business for us to determine the positions of these 'Dark Suns' and to project them onto the zodiac: what we get is a series of points, one for each planet, to which that planet's energy is directed, but where it is not mixed with the influence of other planets, as is the case with the Sun.

In the past, this second focus has been ignored by astronomers and astrologers alike as having no practical part in their reckonings: the former have not found it necessary to explain their laws of gravity and motion, while the latter seem to have lost track of its usage long ago in Egyptian science.

But, in fact, the whole matter of gravity still baffles us in its finer points. Quite why the planets stay in their ellipses if there is no body at the second focus is hard to explain, and as recently as January 1986 an American scientist, Ephraim Fischbach, alarmed his fraternity with the claim that he had discovered a 'fifth force' in nature, accounting for the anomaly by which two bodies of different weight do *not* fall at the same rate as each other. Galileo, it seems, was wrong. Fischbach also shows that different bodies respond differently according to their chemical composition as well as to their mass, so that gravity is far more complex than it was previously thought to be.

Now there is a strong correlation between Fischbach's work and the Dark Suns: a free-falling body on Earth is describing an ellipse, one focus of which is the Earth itself. And if that were all there were to the matter, then Galileo would be right; Aristotle and Fischbach would have nothing to say to us. But Fischbach has made it look as though Aristotle was not so wrong after all, and may have been hazily aware that someone before him already knew what he had just rediscovered in 1986. Each and every body on Earth has a common focus, the planet itself, but then has another which is its own unique property.

As below, so above, to invert the astrologer's maxim: the discovery of elliptical orbits by Kepler, and their refinement and confirmation by Newton, are fine as far as the two men went, but are not in fact explicable in themselves if the second focus plays no part in the matter — if each planet's individual and unique contribution to the system is not counted alongside its role in common with the others as a satellite of the Sun. To draw a parallel from the terrestrial rather than the celestial, how many political dictators have tried and failed to set up a system based on simplistic principles in which humans are just satellites of the state, with no individual and characteristic contribution of their own to make?

That word 'contribution' is crucial to the understanding of the Dark Suns. The whole universe is an interchange of energies, a game of give and take in which the stars give out energy, black holes collect it, and those in between change its forms and qualities as they first receive it and then pass it along. Our local family of planets all collect energy from the Sun; they are all transmuting it according to their individual natures, and then contributing what they have done with it back to the system again. What they receive is all one kind from one source; what they give back is many different kinds, collected at just as many different points, each suited to what it receives from

its attached planet. The solar system is like a factory producing many
different wares from its primary source of raw material, wares as
different as books from a publisher and milk from a dairy-farm, and
the books are not issued to grocer's shops, nor the milk to libraries.

We must now consider what each planet does contribute from its
own work and nature, and how that matters to us here on Earth.
We can best begin with Earth's own Dark Sun and what we know
of it in our history.

The Egyptians paid considerable attention to the Dark Sun, and
the Great Pyramid, the most obvious relic of their culture, provides
us with a clue. The pyramid has often been called 'the centre of the
Earth', and yet it cannot be so: it is neither at the pole nor at the
equator. Then we discover that a few thousand years ago it was at
the place where, at the moment of the Summer Solstice, the centre
of the Earth was in alignment with the centre of Earth's Dark Sun,
and with the centre of the Sun. The diagram shows this situation,
in which the Pyramid was briefly, but truly, the centre spot on this
planet. The centre today is some 300 miles away in south-west Egypt,
and the moment of alignment has shifted some 13 days from the
solstice, which no longer coincides with the Earth being at its
aphelion.

Records of just what took place in the Great Pyramid at this triple
conjunction are now very sketchy indeed but many mystic schools
have long held that the Egyptians used to hold a ceremony there,
in which they ritually transmitted all that they had learned and
discovered during the past year to the Dark Sun, in deliberate
payment for all that they had received from the Sun by way of life-
giving energy. Some Egyptologists also claim to have found evidence
from artistry in the pyramid itself that this happened in the days
when the structure still had its gold or crystal cap on it. There is no

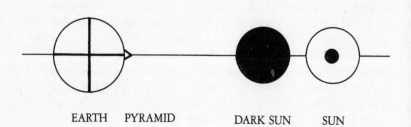

EARTH   PYRAMID   DARK SUN   SUN

evidence that any kind of sacrifice other than that of garnered intelligence took place there then, but the bloodier practices carried out atop other, mostly stepped, pyramids around the tropical world may have had their origins in a similar ceremony.

We could speculate wildly upon how the practice began. Some accounts have the Great Pyramid already in existence before the Phaethon disaster, in which case it was then in the wrong place for 'centre'. It seems likely, then, that it was built in the aftermath of the event (though by whom and by what means is still an interesting question), and the purpose may have had something to do with a determination that never again would the Sun be able to draw willy-nilly upon the resources of its companions; better to pay the account on consciously arranged terms. The cessation of the ceremony, and the eventual removal of the cap of the pyramid, seem to say that a time came when either the sense of obligation or fear of further celestial reprisal wore off, or when the intelligence level on (and off) this planet rose so high that no one particular day and place was needed for the transmission.

One thing is for sure: this planet Earth's unique contribution to the solar system is the intelligence and learning gained from it. Our probes to other planets have now shown us beyond doubt that Earth alone supports a life form which devotes itself enthusiastically to enquiring, learning, and reasoning about what it has found.

For the past 700 years the alignment has been in Cancer as we see it, in Capricorn as the Sun sees it. It has been in the second decanate of Cancer since about 1850, the Scorpionic decanate for us, the Taurean for the Sun. So, our Lunar propensity for learning, which is intuitive, romantic, and religion-dominated, has given place in some measure to the harder, more materialistic, and purpose-oriented kind more suited to an era of industrial growth and a Mars-Pluto sub-rulership. Not that we don't look back, of course, over our shoulders to the older, Lunar shadow which still holds overall sway in Cancer; we can reckon on some 350 years yet before the Piscean decanate makes itself felt, whatever that may bring. From the Sun's point of view, or from the Dark Sun's point of view, this period has been one of gaining hard Capricornian fact from Earth each year at the triple alignment, garnished since 1850 with Taurean usefulness in practical application. The reader may have a good time calculating the position of the Dark Sun in previous eras (it takes 18,000 years to complete one full orbit) and what was being contributed. Just after the Phaethon disaster our Dark Sun was in Capricorn as we see it, and we certainly had a dismal report to make then. From the centre, however, it was in Cancer, and this was certainly the period when life here developed what we now know as an emotional system in our biology.

So Earth's Dark Sun is all about gathering knowledge, applying it, growing intelligence from it, and supplying this gain to the solar system as our contribution to its evolution as a whole. It should not be seen as a solely human activity, though we must accept responsibility for the lead role. The animals we share life with, and especially those we make our servants, pets, and protected species, are all gaining from their contact with us and they must add no small part of the total contribution. So also must the latest advances in our computer technology, where, with whatever limitations, we have set in motion circuits and pieces of silicon that can calculate while we look on.

Where we find Earth's Dark Sun on our individual horoscopes, we find how we are making our unique contribution in this great work. It will be weaker, and more difficult to see clearly, when the Sun blocks our view of it, and stronger when it is between us and the Sun, especially in mid-summer. As always, the traditional language of astrology is best when trying to understand the relevance of the Dark Sun to each of us. In other words, we must look at it through sign, house, and aspect, as we would the Sun.

6.

USING THE DARK SUNS

Bernard Fitzwalter

One of the most useful things an astrologer can develop is a mental idea of the positions of the planets. There is a lot of calculation involved in the erection of a horoscope, and unless the astrologer has quite a good idea of where he expects a planet to be, slips in calculation are likely to remain undetected. It is, of course, one of astrology's most uncanny features that any interpretation done with a chart which is believed to be correct will *be* correct until the chart is found to be in error, but we will pass that by.

We still feel that it is useful to know that retrogression of an outer planet can only occur when it is on the other side of the Earth from the Sun, and that an apparent retrograde conjunction of Jupiter and the Sun means that you have accidentally turned over two pages of the ephemeris at once.

Where do the Dark Suns occur in a horoscope? What are their movements, their likely positions?

The quick answer, of course, is that they don't move, but we do. We are orbiting our Dark Sun as surely as we orbit the Sun itself; that means that as the Earth goes round the Sun, and the Sun from a geocentric viewpoint moves along the ecliptic, so the Dark Sun moves around the ecliptic as well.

There will be times of the year when the Sun and the Dark Sun are conjunct, from our point of view, and times when they are some distance apart. The time of maximum difference, so to speak, is when the Earth is roughly in square aspect to them both, which is a few days after the Equinoxes. The speed of the Dark Sun against the zodiac as seen from Earth will not, therefore, be constant, but then neither is that of the Sun. Taken over the whole cycle, though, they are identical, and in a sense inverted: when the Earth is at aphelion to the Sun, it is at perihelion to the Dark Sun, and vice versa.

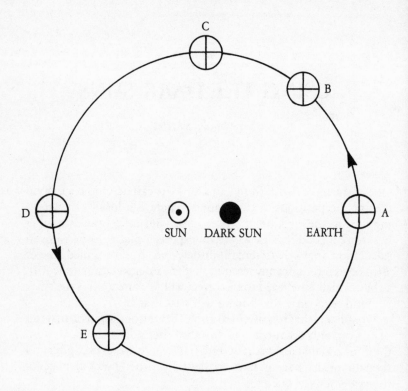

The diagram above will clarify matters. Although the foci of the ellipse as shown are much closer together than in the earlier diagram, they are still much further apart than is really the case, but they will serve to illustrate the point. When the Earth is at point A, the two foci are conjunct from this point of view, with the Dark Sun in inferior conjunction. This happens when the Earth is at aphelion, its furthest point from the Sun. The date for this each year is around 4 July, but whether this has anything to do with the horoscope, or indeed the national character, of the United States of America, has not yet been established!

As the Earth continues its orbit to point B, it will be seen that from a geocentric point of view the Sun has moved against the zodiac. It will also be seen that the Dark Sun has apparently moved further; the difference is entirely due to the difference in space between the two foci, and their apparent speeds are a product of the Earth's motion and the perspective of a geocentric observer. Not that this invalidates the astrology in any way: symbolic values applied to *apparent* positions, geocentrically viewed, are what makes astrology.

The Dark Sun continues to extend its lead on the Sun until the Earth is at point C in its orbit. The Earth is just over 88° from its aphelion, and square to the Dark Sun, with the distance between the two foci now 1°54′50″. From this point on, the angular difference between the two bodies diminishes.

It is interesting that point C occurs with the Sun at about 9 Libra, and the Dark Sun at about 11; this is the same area of the sky as the fixed star Vindemiatrix. The star's name, meaning 'grape-gatherer', was apparently given because grapes were picked to be made into wine at the time when it rose with the Sun. Might it not also represent a time when the 'growing season' between the two foci is over? Does it not also represent the beginning of a time of reckoning, when what was sown is reaped, and when (as the Sun does to the Dark Sun) action and consequence catch each other up? Certainly it would seem that the good times, both of the year in terms of its heat, and in wider terms too, are over from this point on; is this how the star gets its modern nickname 'the widowmaker', because increase is denied hereafter?

Point D represents perihelion, occurring on 4 January or thereabouts. The angular difference between the two foci is again zero, and this time from the Earth's point of view the Dark Sun is in superior conjunction. This is as far as the Earth gets from the Dark Sun, and astrologically our view of it is eclipsed by the presence of the Sun itself: a heightened sense of self-interest could be suggested for those born at this time.

As the Earth completes its orbit, passing through point E, the Dark Sun appears to trail the radiant Sun by the same amount as it had previously led by. Once again the maximum separation comes when the Earth is about square to the two bodies on 3 April or thereabouts, and after that they draw closer from the Earth's point of view until they are once more in conjunction at aphelion.

Putting the Earth's Dark Sun on a horoscope is easy, because it is always going to be within a couple of degrees of the Sun, but the Dark Suns of the other planets are not so straightforward. As the distances from the Sun increase, the ellipses of the orbits get bigger, and the distances between the two foci increase accordingly. Add to that the fact that the observer on the Earth is on a different orbit completely, and the angular displacements can be quite considerable, as will be examined later.

An immediate problem is that of how to represent these points: what symbols do we use?

We decided to 'fill in' the circle of the Sun's symbol, so that it would no longer seem bright and radiant — ●. If we had filled in the circle-and-cross symbol of the Earth, we would have arrived at the same result, so it seemed that our decision was made for us. It is possible that some confusion might occur with the symbol for a New Moon, but the Dark Sun symbol seemed so natural and obvious that it would be hard to ignore it. It might be reasonable to leave a white spot at the centre ◉ and so produce a 'negative' image of the Sun, but that might be open to misinterpretation.

The other planets' glyphs follow on without much difficulty. Mercury's Dark Sun is unique to Mercury, of course; it is no part of the Earth's orbit, and must be shown to be something essentially of Mercury. Filling in the 'circle of the spirit' in Mercury's glyph makes symbolic sense, and is easy to recognize: ☿. Venus ♀ and Mars ♂ work in the same way.

The three modern outer planets are very easy indeed, since their glyphs all contain a convenient circle to begin with: Uranus ♅; Neptune ♆; Pluto ♇. Pluto works best in his 'basketballer' form; the 'ꝓ' version with the bowl filled in looks lumpy and doesn't have the symbolic sense of the other.

Jupiter and Saturn present a problem. Eventually we decided to use ♃ , ♄ respectively, going back to the original message of the glyphs: Jupiter has spirit above matter (circle above cross), while Saturn places matter above spirit.

Chiron is easy enough: the circle below the 'key' motif gets filled in ⚷. The Dark Sun of the position of the ghost of Phaethon, as it now is, has not been determined, and may not be useful; Phaethon and the Sun orbited each other originally, in a much more complex system than the relatively simple ellipses we have been considering.

7.

INTERPRETING THE DARK SUNS

Assigning interpretative or symbolic values to the Dark Suns is no easy task. They must, of necessity, each represent something different, because they are part of the orbital machinery of different planets. At the same time, however, they must have something in common, because they all perform the same function: they are the focus that the radiant Sun is not.

To a certain extent the job has been done for us. At least, the limits within which we can assign meaning are fairly closely prescribed. We already have an accepted body of connotations associated with the Sun, and what we take it to signify. We also have similar idea-networks attached to the other planets. What we must do with the Dark Suns, it seems, is to produce either an inverse or a reverse of those qualities which we associate with the Sun (but not an opposite), and use that as the base quality of a Dark Sun. Then we must take the principles of the planets, and in each case produce an amalgamation of the planetary principle and the 'inverse-Sun' qualities to represent the Dark Sun of that planet.

The interpretations which follow are not hard and fast. They are compounded from the authors' ideas about the Sun, its essence, and its inverse, and about the planetary principles. They are, obviously, subjective, but so is all astrology: the astrologer is the person who notes points of correlation and significance between the sky and the earth, and it is his interpretation which makes astrology. The interpretations given here are intended as a starting-point, a stimulus for the imagination and intuition of others who will work with these points. The traditional planets have a couple of thousand years of accumulated practice attached to them — the Dark Suns have some catching up to do!

Earth's Dark Sun

This is the Dark Sun which has the most immediate appeal and relevance, since it is part of our orbit: we are going round it at this very moment. Astrologically, though, it is perhaps less dramatic than some of the others, since it can only be a couple of degrees away from the Sun from our point of view, and is frequently less. Those few degrees can be enough to make a difference, though; it is certainly worth considering.

To provide symbolic values for it, the first stage is to define what it is that we get from the Sun. Whatever it is that the Sun radiates to us here, we absorb it and use it, and then transmit it back towards the Dark Sun when we have finished with it. The Dark Sun is a collecting point, the target at which we aim our own energies.

The Sun gives us heat and light. Connected with them are the ideas of warmth, brightness and optimism. It gives us life — both in a physical sense by providing us with an energy source to fuel our planet, and in a metaphysical sense through the spark of the spirit.

It has been said that the purpose of this planet is to develop knowledge, to produce and make multiply the processes of intelligence. The Sun gives us the raw material — spiritual light — for that, and it is up to us to use it as we choose. It should not be a one-way process: we should not take without contributing to the system ourselves now and again. If the Sun, as a source of knowledge in all its forms, is a lending library, then not only should we not deface the books we borrow, and take them back when we have read them, but we should eventually contribute to the library ourselves. There are many ways of doing this: writing another book ourselves for others to read, or being noteworthy enough for someone to write a book about us. Of course, the analogy should not be pushed too far, but the parallel is there to see.

We make our contribution, offer back what we have learned, or whatever, in the direction of the Earth's Dark Sun. Noting its position in the zodiac, and in the houses of a horoscope, will give important clues about where we should be aiming. For most people, the Earth's Dark Sun will be in the same sign and house as the radiant Sun. This is not insignificant: important messages are often overlooked because they sound simple and everyday. The message is that the best contribution that you can make is to be yourself, as closely as possible. Make the best use of your talents; accept your limitations, and waste no time trying to be that which you are not. It boils down to the old Greek maxim 'know yourself', which originally graced the entrance to the oracle at Delphi.

What the planet needs is variety. If every person were to be the same, then the richness of variety would be lost, and with it the patterns of intelligence and the creative possibilities that go with it. Therefore each person needs to contribute, if possible, a fully developed version of himself — nothing more, nothing less.

It may well be that a natal aspect to the Sun, say a trine of Mars within 2°, is in even closer aspect to the Dark Sun — maybe a few minutes of arc. These aspects are very much worth considering, and show which planetary energies are likely to be of most use to the individual in making his contribution to society as a whole. The nature of the aspect will offer further detail on the way this energy might be best used, and the sign and house placements too, as might be expected. To avoid the all-too-easy practice of 'over-significance', orbs of one degree or less should be used.

It is interesting to consider, for a moment, the situation that arises when someone is born with the Sun at the very end, or the very beginning, of a sign. Depending on the time of year, the Earth's Dark Sun is likely to be at the end of the previous sign or the start of the next; whichever way round is immaterial — the point is that the Sun and the Earth's Dark Sun are in different signs. The inference must be that such people are born with one set of talents and qualities, but can best make their contribution to society and history by being something else. Examples who spring to mind at once are Queen Elizabeth II and Hitler, both of whom have the Sun at 0 Taurus and the Earth's Dark Sun at 29 Aries. The difficulties of being a leader when born to be a farmer must be considerable.

A similar situation must also arise when the Sun is very close to an angle or house cusp; the Earth's Dark Sun may well be on the other side of the line.

Here, too, is the answer for all those who subscribe to 'cusp theory': the people who read magazine horoscopes avidly, but proclaim that though they should be a Piscean, they feel like an Aquarian. Whilst much of this is to do with ingress times varying against the civil calendar from year to year, there may be occasions when the Sun is in one sign, the Earth's Dark Sun is in another, and when there are particularly close aspects to the Dark Sun from other planets in the horoscope at the same time. In some cases this could be the key to why the native feels the way he does.

Is there anything to be learned from the annual dance these two points perform? From Cancer to Capricorn the Dark Sun leads the way, while from Capricorn to Cancer it trails behind a little. It may be that Cancer to Capricorn people make their contribution by being progressive (in a zodiacal sense, leaning in the direction of the next sign), whereas the Capricorn to Cancer folk would do better to be slightly regressive. There is a pleasing 'summer-winter' rhythm to the idea.

Dark Suns Of The Planets

The Dark Suns of the other planets require a little thought before placing them into a horoscope. None of them are part of our orbital machinery: they are not coincident with our path round the Sun, nor are they a focus of it. The planets receive their energies from the Sun, though, as we do, and they return them to their respective Dark Suns, as we do to ours.

We interpret the energies of the planets by projecting their positions onto the zodiac from a geocentric viewpoint, and using those positions as symbols for certain conditions or ideas. When we project the planets' Dark Suns onto the zodiac we get a geocentric interpretation of the returning flow of the energy of those planets. The circuit runs from the radiant Sun to the planet and back via the Dark Sun: a geocentric horoscope shows how we intercept, or break into, the circuit at a particular moment in time, and how we are likely to see it. We see the nature of that planetary energy as it appears to us by the position of the planet itself in the horoscope, and we see its interaction with other planetary energies by any aspects in the chart. We can also see where that energy had its *source* — the Sun — and now we can see its ultimate destination, the Dark Sun belonging to that planet.

When the planetary Dark Sun is in a natal horoscope, the following interpretation becomes possible: *this* part of the native (the planet) shows itself in *this* manner (sign and house of the planet). This is part of the expression of an individual whose essential energy is *this* (Sun's sign and house), and he contributes *this* part of himself to society in *this* way (sign and house of the planet's Dark Sun). In each case the emphasis is on returning that planet's energy to society, giving back after having taken.

The Dark Suns of the outer planets are a considerable distance from the Sun's position, and can often serve to fill otherwise untenanted, yet strangely significant, areas of the horoscope.

Finally, it is worth paying special attention to the times of the year when the Sun is in conjunction with the Dark Suns of the various planets, from a geocentric point of view. At these times the Sun is hidden from us by these foci, or more frequently, they are hidden from us by the Sun: there is a wealth of symbolism inherent in the idea of, say, being unable to see Jupiter's exit for the Sun.

Mercury's Dark Sun

This must have something to do with the contribution of considered intelligence back to society from the individual.

Mercury stands for all of our mental reasoning processes. To understand something, that is to be able to hold a mental picture of something, what it is, how it works, the position it occupies in relationships (both physical and figurative) to everything else associated with it, is a process described by a single word in the astrological vocabulary — Mercury.

When we notice the qualities of anything, and abstract them from the physical experience away into memory, intellect, or even into words, then that is Mercury too; and so is being able to hypothesize about qualities missing from the present experience, or the likely effects of any changes that might occur.

Mercury, the astrological concept, denotes anything that happens in the head rather than in real life, and it stores the components of experience, analysis and extrapolation in the mind for further use, the commonest storage and transmission medium being words, which were designed for exactly this purpose. The analytical side of the process is shown zodiacally through Virgo, and the reconstructed experiences in verbal form through Gemini: the ratiocination in between is common to both.

The solar flow, then, is from experience into intellect; the returning flow is from intellect into experience. The individual offers what he has learned, experienced, and analysed; the position of the Dark Sun shows where he makes his thoughts and opinions available to all.

It would not be surprising to find Mercury's Dark Sun prominent in the horoscopes of teachers or writers, and a little research shows this to be so. Also, the close proximity of this Dark Sun to the radiant Sun means that on many occasions the native will have either Mercury itself or the Sun conjunct to it.

A stronger flavour is found when Mercury's Dark Sun is in close aspect to heavier planets in the zodiac. The horoscope of B. F. Skinner, the behavioural psychologist, has Mercury's Dark Sun square to Pluto from the tenth to the twelfth houses. Those familiar with his reputation and his work with animals may be able to detect astrology's sense of humour at work in the symbols here.

If Mercury's Dark Sun is afflicted, it may be that the individual may often feel that he knows better than everybody else, that his understanding is somehow superior. He may even try to change the accepted way of things to bring them into line with his theories; certainly he is going to attempt to convince others to see things his way, and will be dismissive of the opinions of others, no matter how valid those opinions may be. This trait is present in everyone, of course, just as Mercury's Dark Sun is present in every horoscope; on the occasions when it becomes a disturbing influence there is usually heavy affliction from other planets in the horoscope.

A pair who share the bombastic pattern of Mars sextile Mercury's Dark Sun, but who are otherwise unconnected, are Mussolini and the astronomer Patrick Moore (b. 4 March 1923), whose opposition to all forms of astrology is legendary.

A far more sinister example, is Dr Josef Mengele, who had Mercury's Dark Sun conjunct Phaethon. His obsessive interest in racial characteristics led him to make brutal experiments in genetic surgery; Phaethon, it will be remembered, signifies genetic inheritance among other things.

A final note: followers of Alice Bailey's teachings on esoteric astrology, and others, have long suggested that there is another planet whose orbit lies within that of Mercury. This they have named Vulcan. One of the stated features of Vulcan is that its orbit keeps it to within seven degrees of the Sun, geocentrically viewed. This corresponds to Mercury's Dark Sun. There are further ramifications of this, involving the esoteric role of Mercury and its Dark Sun, but they cannot be explored here.

Venus's Dark Sun
Venus's Dark Sun is very close to the Sun itself, since the orbit of Venus is less elliptical and more nearly circular than that of any other planet. The Dark Sun is only separated from the radiant Sun by 34′ of arc at the most, and so to all intents and purposes they are fully conjunct.

The interpretation of this constant conjunction is easier than it might seem. The essential nature of Venus is to do with relationship, and the forming of bonds and partnerships through points of contact and common interest.

It has been noted before that the Sun's position can stand for where things are given to us, and that of the Dark Sun where we have a chance to give them back. For the two places to be coincident in a relationship is not unexpected: a relationship is a place where give and take are equal and reciprocal — at least, a balanced, Venus-Libra type relationship is. If give and take are not occurring at the same place and to the same extent then the relationship is unbalanced.

One or two of the features of the annual cycle of this Dark Sun repay a little study. The times of maximum separation are when the Sun is in Taurus and Scorpio — signs noted for their possessiveness and jealousy in relationships.

Inferior conjunction with the Sun comes when the Sun is in Leo. Leo is a generous sign: it gives freely, and has no need to receive in return. Superior conjunction, on the other hand, occurs when the Sun is in Aquarius; from the Earth's point of view Venus is prevented from reaching its exit. Accordingly, Aquarians are cool, reserved people, not given to demonstrative shows of individual affection. The symbolism generated by these astronomical phenomena fits neatly with the tradition.

Mars's Dark Sun
Mars's orbit lies outside our own, and so the angular distances between the foci of its orbit as seen from Earth are much larger than those of the inner planets. The maximum separation here is 16°25'; for the first time there is a real possibility of having planets in aspect to a Dark Sun which are completely out of aspect to the radiant Sun.

The position of Mars itself, acting as an agent of the Sun's energy, shows us where we have our energy and drive. The force which is needed to overcome inertia and put something in motion which was previously static, the initial impulse, is Mars shown through its first sign, Aries, while the exercise of power needed to keep an existing situation under control and subject to the will of its controller is Mars shown through its second sign, Scorpio.

If natal Mars shows the source of our strength and energy, and its sign and house shows the manner and location of the most likely exercise of that force, then Mars's Dark Sun must show the final end point to which that force is directed, and its sign and house must show the manner and location in which that end point is found.

Mars is often, and rightly, taken to signify sexual energy. If this is so, then Mars's Dark Sun must show where that energy goes to in its expression. A little experimentation on horoscopes you know well will prove both interesting and entertaining. Try it!

More seriously, Mars's Dark Sun will show results achieved, the final points total for the projects on which Mars's energy was spent. More cynically, it is possible to say that no matter what Mars's energy is spent on, and with the best intentions in the world, it always seems to end up the same way — the way described by Mars's Dark Sun.

Since Mars represents force applied from without (as opposed to Jupiter, which is force welling up from within) to direct, control, penetrate, or wound, then Mars's Dark Sun may show points in life where we are subject to such forces, and undergo them: points of submission, surrender, wounding and the like. Transits to Mars's Dark Sun may prove to be informative in this respect.

There is no reason why these points cannot be used to link one chart to another: Eichmann's Martian Dark Sun is within 6′ of arc of Wiesenthal's descendant. For Eichmann to be Wiesenthal's 'open enemy' (descendant) is easy to see, but Eichmann's Martian Dark Sun is trickier to interpret. A point of surrender? Possibly, but better perhaps is the idea of reckoning, which also comes into the business of this Dark Sun; one's actions are performed through Mars and the radiant Sun, but they must be paid for through its Dark Sun.

Jupiter's Dark Sun

Jupiter's Dark Sun is not so harsh or demanding in its effects as that of Mars. The point itself is capable of making a semi-sextile to the Sun (maximum separation 29°57′), and that means that a full range of other aspects will be available to it, without necessarily involving the Sun in the same pattern. Like Mars's Dark Sun, it 'leads' the Sun in spring and summer, 'trails' it in autumn and winter.

Jupiter is what makes us richer. Everything which increases itself, or is connected with the directions of upwards and outwards, is connected to Jupiter. That can mean money, of course, and other forms of material wealth, especially if they increase in value as time goes by; but there is also a wealth of intangible values, and they are Jupiter's too. Happiness and optimism, confidence and laughter, all are Jupiter's; and so is sheer, glorious, fluky *luck*. Nor must we forget physical increase, which is the planet's influence shown through the body: growing, getting fat, and of course pregnancy, which provides 'increase' in every possible sense, from hormone levels to the extension of the family tree.

Jupiter's Dark Sun must give us an opportunity to provide increase, enrichment, and good fortune for everybody else, at our own expense. We have, after all, received all sorts of good things through our intercepting Jupiter's expansive influence. In a way, we are highwaymen, 'borrowing' good times and good fortune as they come our way. The Dark Sun offers us an amnesty: we get to keep the profits, and we are free from prosecution. All we have to do is give a little back to the system. We are not even banned from further 'borrowing'. Considered in this light, it should be no hardship to distribute our wealth and knowledge among those who can profit from it. It may gall us to think that they are likely to make money out of what we give them for free, but it is not the voice of Jupiter which prompts such thoughts: rather it is that of Saturn.

Jupiter's Dark Sun is a point of generosity, pure and simple. A point of charity, perhaps. Or a point of investment — investing your own wealth and good fortune in the future of humanity. As always, aspects to this point in the horoscope, and its house position, will show where you have most to give, and indeed how you are most likely to choose to give it.

Saturn's Dark Sun

This is a different sort of creature completely. To begin with, the Dark Sun itself is an 'outer planet'. Up to this point, all of the Dark Suns have lain at co-ordinates in space inside the orbit of the Earth. This means that the Earth goes round them annually, after a fashion, and so from a geocentric viewpoint, projecting onto the zodiac as usual, the Dark Suns all appear to make an annual circuit of the zodiac. In addition, the Sun is in inferior conjunction with these Dark Suns once a year, and in superior conjunction six months later. When the Dark Sun of a planet lies outside the Earth's orbit, neither the annual circuit nor the inferior conjunction can happen. What happens instead is that the Dark Sun appears to swing slowly back and forth over a small arc of the zodiac as we make our annual circuit of the Sun. In the case of Saturn, the arc covers most of the second half of the zodiac, from 22 Libra (near its exaltation) to 10 Pisces.

There is a further point to consider: Saturn's Dark Sun may be an 'outer planet', but it is only just so — at one point during the year it is within four million miles of us, which is decidedly intimate on the scale at which the solar system functions. This occurs at what would have been inferior conjunction, had the Dark Sun been a little closer in. As things stand, on 21 June or thereabouts each year the astonishing situation arises whereby the Sun, the Earth's Dark Sun, the Earth, and Saturn's Dark Sun are all in alignment along the Cancer-Capricorn axis and within a degree of arc. The diagram below illustrates this.

EARTH'S DARK SUN   EARTH   SATURN'S DARK SUN

It goes almost without saying that important undertakings planned for this day should be considered very carefully indeed; the undertaking may prove too great a burden for its proponent to manage. As an example, see the horoscope for the invasion of Russia in 1941 shown later in the examples chapter, where this powerful line-up also forms the ascendant-descendant line.

Saturn represents all larger and more senior structures than ourselves, and the limitations of space and time. Usually these are placed upon us from outside, and we must do the best we can to work within our limits.

The Dark Sun of Saturn shows where we can place limits on others. It is where we can make demands and constraints, where we can impose our own structures. At the same time, of course, it is where we can impose constraints and limits on ourselves, as an inverse to having them put on us from outside. The June line-up is informative here: looking from the Earth towards the Sun, the Dark Sun is in the way, suggesting that this is a time to contribute something from ourselves. If we choose not to do this, and turn the other way, then we face Saturn's Dark Sun, which was standing very closely behind us, so to speak. Obviously the cosmos has ways of persuading us which way to go.

Chiron's Dark Sun

The difficulty here is that Chiron is relatively new in astrological practice, and there is still much discussion about what it should be taken to signify. Perhaps the most fruitful line of enquiry is to examine its peculiar orbit, and to suggest interpretative values based on what it appears to be doing in the heavens.

Chiron's orbit is very irregular. At one stage in its orbit it is closer to the Sun than Saturn, and at another it is further out than Uranus. These distances are modified when transferred to the plane of the ecliptic, but the effect is still unmistakable: Chiron appears to commute between the orbits of Saturn and Uranus.

It is quite easy to build a network of figurative values on to Chiron's singular orbit. It bridges the gap between the last of the personal planets, Saturn, and the first of the transpersonal ones. What was heavy, limited and downward-looking under Saturn can become universal, inspired, forward-looking and revolutionary under Uranus, if Chiron is used to link the two. Chiron becomes an elevator and a revealer: it helps transcend the limits of the physical and mortal universe, and offers a pathway to the universal and immortal universe beyond the reach of our senses.

The distance between the foci of Chiron's orbit is very large, as might be expected. Chiron's Dark Sun is further from the Sun than the body of Saturn, though not by much. Somehow it must enable the contribution of elevation.

This is difficult to express: the idea of an individual being able to contribute his energies to the elevation of another outrages Aquarian Age egalitarianism. At its best, it implies elitism, and at its worst, it may contain hubris. Taking the role of the demiurge is always part of dealing with the trans-Saturnian planets, however, and cannot be escaped. Perhaps the easiest way to express Chiron's Dark Sun is to say that it is to do with giving people a hand up, as though climbing over high obstacles together.

Enver Hoxha (b. 16 October 1908) provides an illuminating example. He was the despotic ruler of Albania after World War Two. Under him Albania was isolated from almost all contact with the rest of the world, while his very hard internal régime set out to make Albania into a modern industrialized nation. To what extent Hoxha succeeded in his aims is not for us to judge, but there is no denying the hard and repressive nature of his methods. His own horoscope shows Chiron's Dark Sun conjunct Saturn in the (solar) sixth house — obviously a character who believed in the maxim of all work and no play.

Chiron provides a link whereby the specific may be elevated to the universal; its Dark Sun will provide the reverse, so that a universal idea can be brought to focus on, and reduced to the level of, the specific.

Uranus's Dark Sun

The qualities and influences we receive from the Sun through its agent Uranus are our distinctive idiosyncracies and the touch of genius latent in each one of us in some field or other. The planet which rolls round the Sun on its side (while the other, more respectable, planets stand more or less upright) is the very epitome of our gifts for lateral thinking, pointed humour, and being different for its own sake: through its influence we make ourselves look at life from another angle, and it does us good.

Whenever Uranus intervenes in life, the oddest things happen — sharp reminders to use our initiative, and not to be satisfied with empty repetition of what we did yesterday, last week, or since our fathers were boys.

Some of these reminders are unpleasant. Some are truly disastrous. All of them, however, stimulate us into looking at what we do in a new way, and to use our individual and collective enterprise to make changes and make progress.

Uranus's Dark Sun is where we can put our intelligence back into the system. Accumulated experience, along with our responses to previous Uranian shocks which have forced us to take actions or reconsider our moves, are all used to form new patterns of thinking, and new approaches to old problems. These are delivered to the rest of society as a constructive programme for the future through Uranus's Dark Sun.

This Dark Sun is in the neighbourhood of Mars's orbit, as far as distance from the Sun is concerned; there may be some association to be made between the forcefulness of Mars and the desire to make things change shown by this Dark Sun. Numerological symbolists may be interested to note that this Dark Sun is eleven times as far from the radiant Sun as Mercury's Dark Sun. There is often a link made between Mercury and Uranus in their association with mental processes — one is often spoken of as a 'higher octave' of the other. Uranus is also associated with the number eleven, not only as ruler of the eleventh sign, but also as one more than ten, i.e. something new, beyond the physical framework of the tetractys, or Pythagorean triangle of ten.

Examples of the 'contributory genius' of the Dark Sun of Uranus are easy to find. It is rather nice to find Einstein with his Sun conjunct Uranus's Dark Sun; somehow it's exactly the sort of thing you would hope to be the case.

There are plenty of other examples, too. Rasputin (b. 29 July 1871) has Uranus's Dark Sun in the twelfth house opposite Mars, and Hitler has it in the sixth, sextile to Pluto in the eighth. This Dark Sun seems almost always to make strong, simple aspects of significance in the horoscopes of the famous. And so it should, of course: if they have made their contribution to history, then the focus of that contribution should be easy to see.

Neptune's Dark Sun

Neptune's orbit is almost circular, like Venus's. This means that its two focal points are likely to be very close together, and that is indeed the case — even though Neptune is a very long way from the Sun, its Dark Sun is within the Earth's orbit, and so becomes an 'inner planet', able to make inferior conjunction with the radiant Sun once a year, and to make a circuit of the whole zodiac once a year as well (from our point of view).

This Dark Sun is only a couple of million miles from Jupiter's, which is startlingly close by astronomical standards. It is interesting to see that Jupiter and its recent associate (in ruling Pisces) have their alternative foci at about the same place.

In keeping with something associated with the principle of being vague and nebulous, Neptune's Dark Sun is difficult to pin down. The tables at the back of the book demonstrate the difficulty: it has been impossible to give mean positions of the Dark Sun valid for the whole century as we have done for the others, and this has necessitated a more complicated routine being given instead.

More than anything else, Neptune gives vision. Far beyond the physical constraints of Saturn and the planets inside it, and at this stage in the century further out even than Pluto, which has ducked inside it, it is the window on the universe, the standpoint from which our solar inspiration can look out to the endless stars. It is also a place for listening to the stars, uninterrupted by the noise of the planets' traffic as they thunder overhead; on a terrestrial scale, it is akin to standing on a promontory and listening to the ocean.

Neptune's Dark Sun is a place through which to contribute insight and vision. It is a place to show others what you believe in, but which they might not have considered; what your imagination has produced, but theirs probably couldn't; and whatever is in your dreams which you would like to put to a wider audience.

Pluto's Dark Sun

The orbit of Pluto, like that of Chiron, is irregular. Sometimes Pluto is further out than Neptune, and at other times it is further in. Quite why this should be so is not clear, though the Phaethon disaster discussed earlier may have had something to do with it.

Pluto's Dark Sun is a long way from the radiant Sun, as might be expected with a planet whose orbit is so far from circular; it is, in fact, more or less where Uranus is, but a little further out. As we noticed with Uranus and Mars, there may be a link in connotation between Uranus and Pluto's Dark Sun, in that both will be associated with upheaval and far-reaching change. Other astrologers have made links in associated interpretation between Pluto, Uranus, and Mars, usually stressing their active and initiating role, and suggesting 'octaves' as an esoteric principle of association; it seems that the chain is actually formed by the bodies of the planets and the Dark Suns of the next ones in sequence. This further suggests that distance in space itself can have symbolic meaning, and be divided into active and passive 'zones', giving those qualities to the planets to be found within them. Anyone fancy trying heliocentric astrology plotted in zodiacal longitude and distance from the Sun?

Pluto's orbit suggests that it could, as it were, provide a listening post, as Neptune does, while it lay outside Neptune's orbit, and then bring its catch home to our system when it passed inside Neptune. What it brings home stays in the system forever; the changes it initiates are permanent.

The association of Pluto with mass transformation, and irreversible change, seems reasonable, and widely accepted. The inverse of that, as an interpretation of its Dark Sun, must be to contribute to that transformation in some way, to provide the 'deep change experience' which makes a real difference to the lives of all who come later.

Pluto's Dark Sun is always between 10 and 16 degrees of Taurus, geocentrically viewed; there must be people, Taureans whose Sun is conjunct Pluto's Dark Sun, whose whole existence somehow makes them the key figures in the process of mass transformation. These people must have had personal experience of actually being the spanner in the works, the agent for change after which nothing could ever be quite the same again.

Two examples come to hand almost at once: Sigmund Freud and Karl Marx. Both Taureans, of course, and both with the Sun conjunct Pluto's Dark Sun. Marx's placing is the closer of the two — perhaps his work has had an influence on a greater number of people! Obviously there is more to being a major influence on world thought than being born on 5 May or thereabouts, but the examples are instructive none the less and provide further astrological evidence for the prominence of these two men.

A Note On Retrogression

The Dark Suns of Saturn, Chiron, Uranus and Pluto are capable of retrogression: that is, when viewed from the Earth they appear to be moving backwards against the zodiac at certain times of year. As with other retrograde planets, it may be that the contribution suggested by the particular Dark Sun must be re-examined or repeated before full expression is possible. Stations, and the times when a Dark Sun retrogresses over an original position, or vice versa, should prove to be significant.

It is comforting to consider the fact that the Dark Suns which can move retrograde are incapable of inferior conjunction. Therefore, no matter how pressing their influence, they cannot block our view of the Sun; thus there is always more power at hand for the individual to draw upon when demands are made on him by society.

8.

DARK SUN EXAMPLES

The Order To Invade Russia, 1941
This is the horoscope of an event, not a person. It is set for Berlin, from where the order was issued.

The major feature of this chart is the alignment of the Dark Suns of the Earth and Saturn. This is an annual event in the heavens, but this chart merits particular interest because the horizon of the event is the axis of the alignment.

The horoscope is an odd one in many ways. It may well have been elected, i.e. the moment was chosen to reflect and encourage the aims of attacker, but that seems to produce even more problems. The Sun is sextile to Hitler's, and is about to rise, both of which are good things from his point of view; Mars is at the top of the chart to suggest a warlike enterprise, and in the eleventh house here to denote his military ambitions. The Moon is at closing semi-sextile to the Sun, which is a good position for a surprise attack.

It is difficult to believe that nobody noticed the opposition of Mars and Neptune, showing the dissipation of military force as the ever-widening front made a small army struggle to command the enormous expanse of Russia; even if Neptune is ignored and interpretation is limited to the traditional seven planets, it seems foolhardy to declare war when the ruler of the midheaven (the victory) and the ruler of the descendant (the enemy) are the same planet! The inference must be one of playing into the hands of the enemy.

The inference is exactly the same when the Dark Suns are added. Opening an offensive at this particular time seems to be a self-imposed burden on the attackers, and they are forced to contribute their strength to their enemies (Saturn's Dark Sun on the descendant). A glance at the fourth house will show how they fare in the long run: the Sun rules the fourth, but the attackers cannot see the Sun

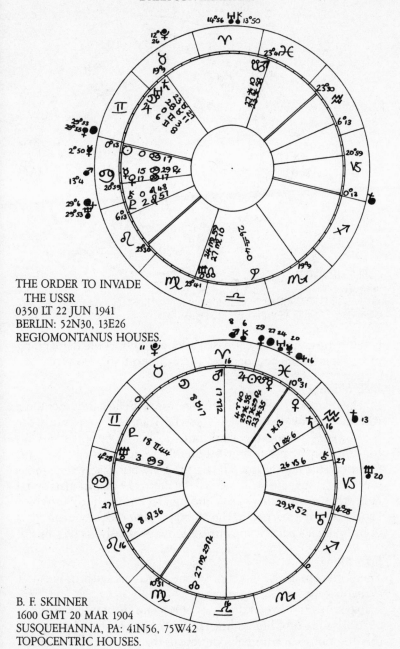

THE ORDER TO INVADE
  THE USSR
0350 LT 22 JUN 1941
BERLIN: 52N30, 13E26
REGIOMONTANUS HOUSES.

B. F. SKINNER
1600 GMT 20 MAR 1904
SUSQUEHANNA, PA: 41N56, 75W42
TOPOCENTRIC HOUSES.

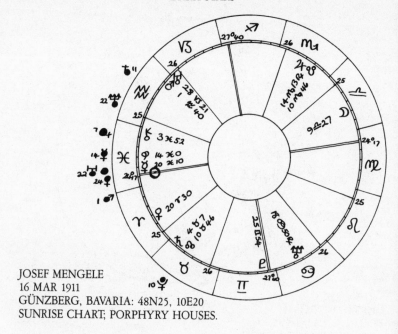

JOSEF MENGELE
16 MAR 1911
GÜNZBERG, BAVARIA: 48N25, 10E20
SUNRISE CHART; PORPHYRY HOUSES.

because the Earth's Dark Sun is in inferior conjunction, demanding more expenditure and denying them support and recovery.

B. F. Skinner

Burthus Skinner's theories of behaviour are a useful antidote to those whose views on man and his psyche lean too far towards the Neptunian. Adding the Dark Suns to his horoscope shows his contribution to psychology admirably well, and they catch his unmistakable flavour too. The Earth's Dark Sun is very close to his south node, suggesting interests lying down the evolutionary spiral rather than up. Mercury's Dark Sun squares Pluto, while Pluto's own Dark Sun is approached by the Moon — what the public considers to be Skinner's particular contribution is obviously his pessimism!

Josef Mengele

The horoscope given is a sunrise one, since his birth time is not available to us.

Experimentation on human victims seems incomprehensible in a sensitive Piscean, but adding the Dark Suns to the horoscope of this notorious war criminal soon reveals the peculiar ways in which he felt he could help the race improve itself.

The Dark Suns of Uranus and the Earth are conjunct, and within a degree or so of the Sun: obviously his view of the ordering of things was radically different from anything that had gone before. It is a configuration he shares with Einstein, the next example. What Einstein lacked, mercifully, was the desire to make quantum leaps up the scale of biological evolution shown here by Pluto's Dark Sun on the north node. He did not have the strange conjunction of Mercury's Dark Sun with Phaethon either; this gave Mengele his opinion that his own views about genetics were the only right ones, and that the world needed to be shown the way.

Einstein

As in the previous example, there is the configuration of Uranus's Dark Sun with that of the Earth, and with the radiant Sun itself. The darker contacts which characterized Mengele's chart are absent here, and in their place is the wonderful interplay of Jupiter, Neptune, and their respective Dark Suns. These points are all in good and creative aspect to one another, and seem to show imagination and mental creativity in a very fruitful form. It is also, thanks to the Dark Suns, what Einstein put back into the system: he may have been a great many things, but it was his capacity to envisage the mechanics of the universe which was his lasting contribution.

Mother Teresa of Calcutta

Whichever way this horoscope is approached, the same message shines from it: self-dedication.

Mars's Dark Sun is approaching superior conjunction with the radiant Sun, having just conjoined Earth's Dark Sun. This happens each year when the Sun is at 5 Virgo, with the inferior conjunction similarly at 5 Pisces. The emphasis on Virgo and Pisces predisposes interpretation towards the dedication of the self to the service of others, but the symbolism of the conjunction in no way contradicts it — superior conjunction means that the goal of one's actions must be through, and beyond, the self. It is as simple as that.

To reinforce this idea, Mercury's Dark Sun is conjunct Mars. Mercury rules both of the angles in this chart; therefore the contribution of the whole person and their life (the angles) is given back through works and deeds (rising Mars), the aim of which (its Dark Sun) is to go beyond the self (superior conjunction with the

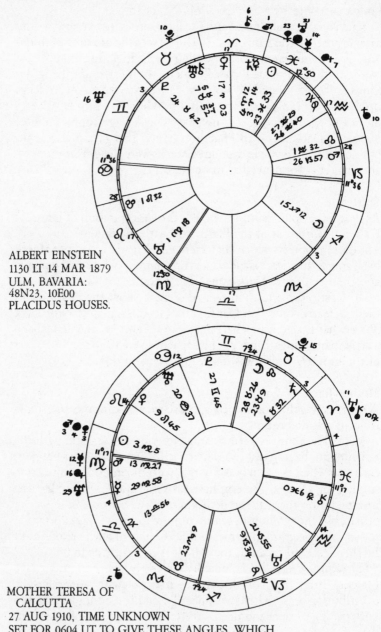

ALBERT EINSTEIN
1130 LT 14 MAR 1879
ULM, BAVARIA:
48N23, 10E00
PLACIDUS HOUSES.

MOTHER TERESA OF
    CALCUTTA
27 AUG 1910, TIME UNKNOWN
SET FOR 0604 UT TO GIVE THESE ANGLES, WHICH
ARE FELT BY MANY ASTROLOGERS TO HAVE VALIDITY.
TOPOCENTRIC HOUSES

Sun) in giving (Earth's Dark Sun . . .) service to others (. . . and the radiant Sun in Virgo in the twelfth house).

Mars is crucial to understanding the horoscope; only through action and work can she give herself fully — prayer and contemplation are not enough. Something of her calling to be a nun can be seen in the opposition of Saturn's Dark Sun to Saturn itself, in the ninth: she was willing to impose the limits and restrictions of a formal religious life on herself, as a framework for her daily life (the Dark Sun is in the third).

Finally, we find that Mercury, the ruler of her angles, is conjunct the Dark Sun of Neptune. She is, as has often been said, an inspiration to us all.

PART THREE:
PLANETARY NODES

9.
ASTRONOMY AND ASTROLOGY

It is a curious thing that in astrology the nodes of the planets are all but ignored, while the nodes of the Moon are given great prominence. How has this odd state of affairs come into being? Is it because the interpretation of the Moon's nodes is easy, while that of the planetary nodes is obscure? This is possible, but it is not likely. The values assigned to the Moon's nodes to express their 'nodeness' rather than their 'planetness' can be transferred to the other nodes just as well, and understanding of the whole meaning of nodes is increased thereby, so interpretation should be easy.

Interpretation of the Moon's nodes is *not* easy, though. There are many astrologers who find the whole question of the nodes a very grey area indeed. They are not planets, and yet they must be important, so just what is it that they do? What is the nature of 'nodeness'?

The origins of this state of uncertainty lie in the modern approach to astrology which, though it has countless benefits, has a few drawbacks too. In centuries past the student learned his astrology and his astronomy together, and his mathematics too; nowadays we learn our mathematics at school, and our astrology a lot later, usually without astronomy. Many astrologers like to distance themselves from the mathematical side of things: it is precisely because astrology is *not* part of the world of modern science that they are drawn to it. In addition to that, modern astrology has gained a great deal from its contact with psychology. The result is a much softer, Piscean sort of astrology, where impressions and meanings can flow and blend easily, and where the personal viewpoint is most readily expressed. The casualty of this approach is Virgoan astrology, where the rules are adhered to, and the individual submits himself to the laborious processes of detailed learning to obtain a clearer vision. As with all oppositions, to add emphasis to one end puts the other at a disadvantage — a heavy child on one end of a see-saw leaves his companion stranded and unable to move.

As astrology has moved further and further towards the Piscean, increasing its understanding of the subconscious and unconscious mind on the way, it has produced ever fewer astrologers who have found it either desirable or productive to familiarize themselves with the mathematics of astronomy, and to seek astrological meaning there.

A further complication is that the planetary nodes are not usually in the ephemeris. It is probably true to say that the contents of the ephemeris exercise a greater influence on the practice of astrology and the techniques it chooses to accept or reject than anything else. The popularity of the Placidus house system rests entirely on the fact that it has been at the back of *Raphael's Ephemeris* for as long as anyone can remember, and its relatively rare use in Europe has more to do with the distribution network of the publishers of Raphael's than with anything else. On a more positive note, the inclusion of Chiron in the *American Ephemeris* so soon after its discovery has meant that this little planet has found a place in everyone's astrology much more rapidly than would otherwise have been expected.

The reason for the absence of the planetary nodes from mainstream practice is simply unfamiliarity. Uncertainty about interpretation, the absence of their positions from the ephemeris, and unfamiliarity with the mathematics required to locate them conspire together to exclude them from regular use.

All that is needed is for the nature of nodes, their positions in the sky, and their meaning, to be made a more frequent feature in the interpretation of horoscopes. Practice and familiarity will soon bring them into regular use, and so complete the symbolic vocabulary necessary to make sense of life on the third planet: the positions of the other planets, their nodes as their orbits intersect ours, and their alternative foci (Dark Suns) as we cross the axes of *their* orbits.

Planetary Nodes — Astronomically

A node is where the plane of a planet's orbit, not the planet itself, intersects the plane of another planet's orbit. The diagram below makes the situation clearer. From the point of view of someone living on the Earth, the Sun's movement against the stars through the year sweeps out a circle, and with a bit of imagination this can be seen as a disc, the plane of the orbit, with the Earth at the centre. The radius of the disc is of course infinite, but to make things simpler the stars are all on the inside surface of a celestial sphere, which is easier to imagine.

Any other planet, as seen from Earth, also describes a path round the zodiac, though not necessarily in a year: Jupiter takes about 12 years, and Saturn nearly 30. The orbit of such a planet can also be seen as a disc or plane, centred on the Earth. The two planes are not, however, coincident — they are at a slight angle to each other and must intersect. The points on the celestial sphere where the two orbits intersect are the nodes. When the ecliptic, or Sun's orbit, is crossed by the path of another planet, so that the other planet is then on the north side of the ecliptic, then that is the north, or ascending, node; when the other planet crosses the ecliptic to the south side, then that is the south, or descending, node.

Nodes stay static, more or less. They do move along the ecliptic a small amount as the planets work through their longer cycles, but the amount they move in, say, a century, isn't much more than a degree. They *appear* to move a lot more than that, though, and that is because of the Earth's movement round the Sun during the year. The nodes of the inner planets, and Mars, appear to complete a circuit of the zodiac once a year; the nodes of the outer planets appear to move backwards and forwards through arcs ranging from about twenty degrees down to just two or three.

The diagrams on pages 94-5 show why this should be so. In the case of the inner planets, the Earth's own orbit either catches up on or swings away from the planets' nodes, accounting for the unusually rapid change in the positions of Mars's and Venus's nodes in May and November. In the case of the outer planets, the nodes are so far away from the Earth that they seem hardly affected by the Earth's movement during the year. Whether we are here on one side of the Sun, or 186 million miles away on the other, Pluto's north node still appears to be in the middle of Cancer. The distances involved are indeed huge.

Since the position of the observer on Earth during the year makes a difference to the apparent position of the planetary nodes, but the nodes do not in themselves move much, then it follows that they will be in more or less the same place at the same time each year, in the same way that the Sun is. Accordingly, there are tables at the back of the book which give the mean positions of the planetary

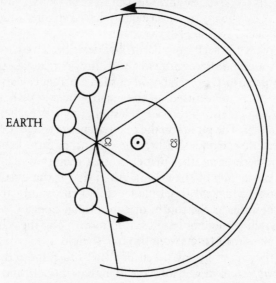

SLOW APPARENT MOVEMENT OF INNER PLANET NODE

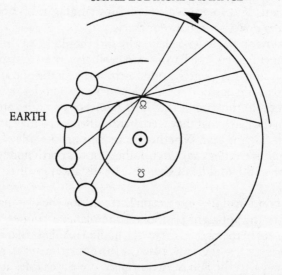

SLOW APPARENT MOVEMENT OF INNER PLANET NODE

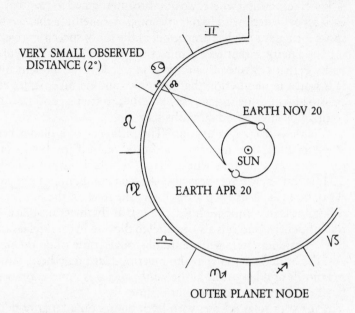

VERY SMALL OBSERVED
DISTANCE (2°)

EARTH NOV 20

SUN

EARTH APR 20

OUTER PLANET NODE

nodes for the first day of each month this century. They are accurate
enough for most purposes — certainly enough to give a good example
of what the planetary nodes have to add to a horoscope. For those
periods of the year when the nodes of Venus and Mars are very rapid
in movement, weekly rather than monthly figures are given.

Planetary Nodes — Astrologically

Any interpretation of the nodes must follow from their physical
attributes. A node is the intersection point of two orbits, as seen
from Earth. Influences meet and cross at these points, and then
proceed in new directions: each goes north or south compared to
the other one, and differently to what it was before.

A node is formed from the meeting of two paths; it is the meeting,
the coming together in time and space, which gives nodes their special
quality. They correspond to places and moments where the path
of things can change, take new direction, or meet another influence.
Such things, when they occur in human life unforeseen, are termed
'fated'. It is a pity that we do not use the word 'node' in describing
these things, since it is beautifully suited to the job; its root meaning

is 'knot', something where two lines are intertwined so that they join each other. 'Fated', with its root meaning of something that has been spoken, suggests predetermination, or the use of speech as a creative tool by a deity; either way, it misses the idea of the meeting of two paths. Perhaps 'crossroads' is the best we can do in the modern idiom. It is worth remembering that not only can one meet strangers at crossroads, but one can also say goodbye to friends and travelling companions as individual paths part.

A node is like a conjunction. The influence of a planet lies all along its track, like the stripe you would paint if you held a brush to a vase on a potter's wheel. It is this which gives parallels of declination their force, and also the paranatellonta of the astro-cartographers. When a planet crosses the track of the Sun, it is as though it is in conjunction with it: it is at the meeting-place, and the influences join. It is also parallel to the Sun in celestial latitude. It is at its node. There will eventually come a time when the planet and the Sun are actually at the meeting-place together, conjunct in latitude and longitude at the same time; this super-conjunction is, of course, an occultation, or eclipse.

A lunar or solar eclipse, with both bodies on a lunar node, is a force of such power that an event is almost bound to happen, as experience has shown. It can even be specified to quite narrow limits in time. The natural forces are so strong in mutual support as to virtually prohibit any attempt at circumvention or manipulation — the current is too strong for us to swim upstream. If the current is flowing in the desired direction of travel, then progress will be rapid; otherwise, by far the best course of action is to wait for the surge to abate.

But what if neither Sun nor Moon is on that node? What if only one of them, or some other planet, is? Or if nothing is on the node, but simply makes an aspect to it, or conjoins it by purely symbolic progression?

An example may help. Two men are standing on a cliff edge. One of them has his Mars on the day's south lunar node, and, careless of where he is, he steps forward. He at once applies all the planetary forces (Moon being the common amplifier of them all) and descends to Earth very swiftly. The other has his Saturn square to this node; he stays rooted to the spot in fear until, overcome by vertigo, he topples after his foolhardy companion.

Now a third man steps up, having his Mercury sextile to the node in question. It is the south node, so whatever else he was intending to do, he feels he must stop and help. He listens for a cry from below, hears one, calls to others to fetch the coastguard, and remains to mark the spot for them.

There was no eclipse or occultation here; the Sun was not engaged, nor was the Moon, but all three men reacted as though it were the real thing, an eclipse, and so they 'made' one there.

It is this peculiar quality of nodes which has caused many astrologers to describe them as 'spiritual'. This may be in the hope that someone else will understand them better than they do themselves, but it may also be an interpretation of a point in the zodiac which has no physical object, such as a planet, to mark its place. This is faint-hearted astrology: the zodiac itself has no more reality than the nodes, nor do the houses, or the angles. Yet we use them as though they were as solid as the planets — we should have no fears when handling the nodes.

What makes a node different from, say, a house cusp, is that it has a specific *kind* of non-physical nature. It has the memory of the paths of two planets. It marks one of only two places in the sky where they can meet exactly (from the viewpoint of someone on Earth, which must never be forgotten). What is more, it marks not only where those planets have been, but where they will meet again. Nodes are the very emblem of cyclical time: a place where events once took place and where they will again.

What a wealth of associations can now be attached to the nodes, once the idea of time is represented in them! The south node becomes the past, looking back, and the north node becomes the future, waiting for the event it knows will come. It is easy to see how the nodes of the Moon are associated with 'karma', as we abuse the term in the West.

Why is it that the nodes of the Moon are the ones which are used so often? Is it that the Moon's nodes are more influential than those of the other planets?

Probably not. There are a number of possible reasons, all of which make good symbolic sense:

(a) The Moon is concerned with the inner, reflective side of the individual. It must also be concerned with his soul, since the other light governs his visible self. Any philosophy which embraces reincarnation would look to the Moon's nodes rather than those of any other planet to see the paths the soul has already been on, and those it has still to travel.

(b) The Moon is the innermost sphere in the philosophical arrangement of the universe. It is therefore the final link in the chain of causation from the higher spheres down to Earth. Its nodes must be where it contacts the rest of the celestial clockwork, and so to be fully aware of coming events it is prudent for the astrologer to take note of their position and movements.

(c) The Moon is the body whose nodal cycles cause the most striking effects when viewed from Earth, i.e. solar and lunar eclipses. Since the Sun is the focus and source of energy for the whole planetary system it is reasonable for the astrologer to be aware of the cycles which interrupt the flow of that energy to Earth.

(d) The Moon's nodes are in the ephemeris, and have a special pair of glyphs which are dutifully learned along with those for the rest of the planets. Astrologers, like fishermen and priests, are the most conservative workers in the world; they resist innovation to the last, and even when they accept it, they do not let their old practices lapse, for fear that their luck should change.

There is an unfortunate feature about lunar nodes which has always made their interpretation difficult. This is that they are in opposition to each other, and so both are effective at the same time. They are also both in similar quality of aspect to any third body, except when something is actually conjunct one node or the other. It is therefore often far from easy to tell which node is having the stronger effect: is it the pull from the past, or the impulse to the future?

The planetary nodes are rather easier to use. From a geocentric viewpoint, the nodes of the planets as far out as Jupiter are only occasionally in opposition, so that the past-future polarity is more easily distinguished.

There is a way out of the lunar node dilemma, and that is to count the effective north node to be where the Moon last conjoined it, and the effective south node to be where the Moon will next meet it, or vice versa. Even so, the difference thus engineered is very small, and may not be enough to help in interpretation.

The nodes of each planet may be expected to correspond to events of the nature of that planet. When some planet is at another planet's node, we should expect to see the effects of a triple conjunction, as though the first planet, the planet whose track is here crossed, and the Sun, were all present. The event should have a past or future leaning, according to the node involved, and perhaps it will look backwards or forwards to the next or last time the planet itself was on its node, when an astrologically parallel and similar event may occur (or have occurred).

The effect of strong nodal configurations is to stimulate the native to actions of which, by the showing of his ordinary — planetary — horoscope, we would not have thought him capable. What happens is that he comes to a crossroads, meets his fate, and makes his choice. It is simply a question of being in the right place at the right time, and it happens to the most unlikely people, as the examples will show.

10.

INTERPRETING THE NODES

Interpreting the planetary nodes is slightly different from the usual sort of interpretation. What is required is some sort of modification of the values associated with the planets themselves. The nature of the planet must be there for all to see, yet there must be something extra, something distinctly nodal, to distinguish it from the planet itself. In giving interpretations of a planet and its nodes, we must make sure that we do not say the same thing three times, and yet we must also make sure that there is a definite and clear link between all three points and the way they show themselves in the horoscope.

In addition to the purely planetary quality of the nodes, the signs and houses they occupy in an individual horoscope should also contribute to their overall meaning, as is the case with the planets themselves.

When interpreting the significance of a planet in a horoscope, there are three major factors to consider: the nature of the planet, its sign, and its house. With a planet's node, on the other hand, there are *five* major factors: the nature of the planet, the general nature of nodes, the polarity of the node (i.e. whether it is a north or a south node), its sign, and its house. Obviously this produces a piece of symbolic vocabulary capable of representing very complex or delicate nuances of meaning, but at the same time one which requires practised handling if its subtleties are to be fully brought out.

In no case, however, will the planetary nodes give any indication of any quality or event which is not already visible in the traditional, planets-only horoscope. What they have to offer is a reinforcement of the symbolism of the planets themselves, or a way of showing the same event from a different viewpoint. Whilst this adds depth of focus, it does not bring anything new into the picture, and to suggest that the nodes offer essential information which the planets themselves cannot supply is false astrology.

It was suggested earlier that the nodes lent themselves to an association with the idea of time, and especially that of cyclical time. It was also suggested that since nodes are the points where orbits cross, they could be taken to represent points in time and space where an opportunity occurs for a change of direction. A good way of visualizing this is as though nodes were the switching points on a railway line, where the train is enabled to take one track or another according to the way the points are set.

Like real railway points, or real forks in the road, nodes represent choice. Which road is taken is up to the traveller. He can take whichever road pleases him, but he cannot take both at once — he must make some kind of a choice. In the case of a north node, he can choose to maintain his present course, or change to a route which will take him not only along but up as well.

Nodes can be seen to demonstrate some of that element of free will which astrology's detractors seem unable to find in it. The choice is entirely open; opportunities presented may be taken, or they may be passed by. There are consequences which follow either action, of course, but the choice is a free one all the same.

If all of these strands of thought are brought together, then the north node must stand for a point in time which provides an opportunity to use the energies of a particular planet to go up and forwards with a view to the future. It seems a reasonable hypothesis, but some of its phrasing hides a wealth of tacit assumption.

The word 'up' is a case in point: the assumption has long been that 'up' and 'north' are equivalent to 'good'. To suggest that up is good is astrologically defensible, but it is difficult to disentangle it from any political implications in the modern mind. Astrology is not necessarily elitist, but it does believe that progress from Earth to the heavens is desirable. There is a geographical element to be considered, too; for a traditional astrologer in the northern hemisphere, where, as history would have it, he has usually been, the chain of association *up — north — pole — heaven — God — good* is both reasonable and easy to understand.

The implication seems to be that a planet's north node offers an opportunity to make a step forwards, to become involved in something new, where the energies of the planet concerned can be expressed in a way which will bring benefit to the native. If the call is ignored, then the opportunity passes by, and that particular development and growth does not take place in the individual.

South nodes appear to be concerned with giving back rather than receiving. The benefit which the individual has accumulated by taking north node paths when they are offered is returned, in a modified form, by the south node paths. Throughout this book there has been an emphasis on the 'returning circuit' in astrology, where energies are received, modified by use, and returned to the system, so that the cycles can continue. South nodes are simply the returning circuit of the north nodes.

South nodes represent a point in time and space, a situation, where things are demanded rather than offered. Rather than being an opportunity to take extra wealth into the life, they offer a chance to unload, by contributing previously-gained wealth into the situation at the point where it is needed. If the demands of a south node are ignored, then the situation will overwhelm the native, to his detriment.

The qualities of the nodes are similar, in many ways, to the qualities of the traditionally-termed 'hard' and 'soft' aspects. North nodes, like trines and sextiles, offer and present benefits, and develop talents in profitable directions; south nodes, like squares and oppositions, make demands on time and resource, and force decisions to be taken. It is part of human nature to recognize the situations engendered by harsh aspects and south nodes much more readily than their 'benefic' counterparts, though in the long run it may be seen that to make the best of the opportunity presented by a north node may require more effort than the native appreciates. Conversely, the south node of a planet frequently offers a way of 'releasing' the energy of that planet in a way which is beneficial to all concerned.

Though they may be similar in some respects, planetary nodes are very different from planetary aspects. Aspects represent a static state, a fixed relationship which permeates every level of existence. Planetary nodes represent particular places in time and space, gates through which the native chooses to pass, or not, as he reaches them. They are especially effective in showing how the native chooses to develop his talents, and at which times. Their use in character interpretation is probably limited; their greatest potential use is likely to be in work with progressions and transits, used to delineate opportunities as they arise.

Mercury's Nodes

At first sight these look a little unusual, but their oddness is entirely to do with Mercury's proximity to the Sun. Since Mercury itself is never more than 28° from the Sun, from our point of view, then its nodes can never be separated by more than twice that figure, and will often be much closer.

These nodes can, in fact, appear to be conjunct at certain times of the year, which comes as something of a surprise to those accustomed to working with the perennially opposed pair of the lunar nodes. The interpretation of a nodal conjunction stretches the mind somewhat, especially if one's interpretation of nodes previously had leaned heavily on the fact that they were in opposition to each other, but it is not impossible. At such times, of course, they must also both be conjunct the Sun from a geocentric point of view, since what has in fact happened is that the Earth has reached the intersection point between the planes of Mercury's orbit and its own.

The two dates for the conjunction of Mercury's nodes are around 9 May and 9 November each year, with the two nodes and the Sun at about 17° Taurus and Scorpio respectively. It is probably worth doing a little observation and research to see whether these dates offer minor turning points, transiting opportunities as it were, to those for whom Mercury is significant: Sun Geminis, Sun Virgos, and the very great number of people who in these latitudes are born with Virgo rising and a Gemini MC.

It would seem reasonable to suggest, in interpreting these nodes, that the north node would present opportunities for thought, inquiry, and analysis, with some element of communication, such as public speaking, or writing, associated with it. Rejecting the call of this node would simply be to turn down the chance to take part in this stimulating and expressive experience. In the same way, the south node would represent an occasion when the native was called upon to analyse and consider his situation, and contribute his opinions for everyone else's benefit. On such an occasion his opinions and analysis matter far more than those of anyone else; only he has the correct viewpoint essential for the solution of the problem, and it is up to him to offer it.

Mercury's nodes seem to be significant in cases where the intellect has been applied to the analysis of specific techniques, particularly those involving the body, the mind, and any element of deception. All of these qualities are easily identifiable as Mercury's, through its signs Gemini and Virgo. The charts of Houdini and Sidney Piddington are most informative in this respect, and their data is given in the examples section.

Those born on the dates when the nodes are conjunct are equally interesting. There seems to be a preponderance of poets and composers — Browning, Brahms and Tchaikovsky, for example. Is this similar to the highly creative intellect shown by those born with the Sun and Mercury conjunct? It would appear so.

In interpretation of these points, orbs of about one degree should be used wherever possible. Using a sieve with smaller holes in it usually results in a finer product; to adopt wider orbs is to invite misinterpretation.

Venus's nodes

Venus's nodes share many of the odd characteristics of Mercury's nodes; this is because the mechanics of their orbits are similar in many ways, especially when viewed from Earth. Once again, because Venus is an inner planet, it is possible for both of the nodes to conjoin not only each other but the Sun as well, and this happens twice a year — as was the case with Mercury.

The movements of Venus's nodes are extremely rapid at these times, as a glance at the tables at the back of the book will show. What is actually happening is that the Earth is 'catching up' on the node in question. Towards the end of May, the north node appears to move very slowly compared to its partner. This is because it is on the far side of the Sun from us, and therefore the angle subtended between it and us, viewed against the zodiac, does not change much. The south node, however, appears to be accelerating towards it at quite some rate. In reality it is the Earth which is gaining on the node, but since it is closer to us its apparent movement is much greater. The Earth reaches the intersection point of the two orbital planes on 7 June, and on that day the Sun appears to have reached the north node at the same time as the south node reaches the Sun: all three are conjunct (and so, of course, is the Earth, as in any visible alignment).

The reverse situation, where the north node is the swifter mover, occurs in late November, with the triple conjunction around 8 December, with the Sun at 15° Sagittarius.

The north node of Venus probably represents an opportunity to acquire wealth or to form significant partnerships. It seems to attract good things towards the native, which would seem reasonable. Venus is, after all, the minor benefic, and it is also associated with attraction, as in magnetism. Choosing not to heed the call of Venus's north node must simply mean that the native does not want material success. A good example of Venus's north node at work can be seen in the chart of Howard Hughes, the famous reclusive millionaire, where the node is on the cusp of the second house, in Aquarius. The cusp of the second house must surely indicate his considerable fortune, and the sign of Aquarius is a good indication of the area of his greatest interest: aviation. The famous astrologer Ivy Goldstein-Jacobson has the north node of Venus in the tenth house, conjunct Lilith — entirely appropriate for somebody who practically introduced Lilith to most astrologers through her little book on the subject.

Venus's south node shows where Venus's energy has to be given back. Often this will mean contributing money, but it will also indicate a requirement for reconciliation and affection, the intangibles of personal relationships which are so much a part of Venus's realm. Like giving away money, giving away warmth and affection in an effort to restore harmony and balance requires a mature perspective: many people will find it hard to do. It is important to see the south node not as where things are *taken* away, but where they are *given* away. The separation (an 'untying' of the Venusian bonds of attraction) is the same, but the sense of loss is much reduced.

Mars's nodes
Because the orbit of Mars is further from the Sun than our own, the nodes of Mars appear to swing back and forth about a well-defined axis during the course of the year. The axis is Taurus and Scorpio, though the north node appears to swing between Aries and Gemini, while its counterpart moves between Libra and Sagittarius. Although the two nodes are naturally at opposition, the Earth's position means that they only appear as such on two days per year, when the Sun is conjunct one and in opposition to the other. As might be expected, the movement of the south node in May, and the north node in November, is very rapid; the reasons are similar to those explained above with reference to Venus.

Mars's nodes indicate an opportunity to project the self, and to make a personal contribution to a given situation by sheer force of will and individual personality. Since Mars denotes an intensity of

personal force rather than a particular field of activity, the sign and
house position of these nodes will be especially useful in determining
their most likely expression. Mars's north node in Aries, for example,
gives an opportunity to project the self and take the lead in the initial
stages of some new enterprise, whereas Mars's north node in Taurus
gives an opportunity to project the self in a practical, down-to-earth
way. Similarly, Mars's south node in Aries demands that the native
makes his presence felt, by contributing his literal presence and
qualities of leadership, while Mars's south node in Taurus demands
that he do something constructive, practical, and useful for the benefit
of all. The *way* in which the native chooses to do this may be discerned
from the sign and house position of the natal *planet* Mars: the nodes
show the opportunities for its expression, which is not quite the same
thing.

Mars's nodes and the Sun are in alignment when the Sun is at
19° of Taurus or Scorpio. People born on those days should display
qualities of personal determination and will-power, which, when
coupled with the fixed nature of the signs involved, should bring
them to prominence through sheer persistence. President Truman
('the buck stops here') was one such person, and the Chinese
revolutionary leader Sun Yat-Sen was another. Their birthdays were
six months apart, and so for Truman the Sun was conjunct the north
node, opposed to the south, while for Sun Yat-Sen the Sun was
conjunct the south node and in opposition to the north. Students
of political history may like to mark the similarities and differences
in the two men.

Jupiter's nodes

The nodes of Jupiter, and those of the planets beyond it, do not
appear to move very far from their axis at all; as the distance increases,
the movement lessens, as one might expect.

The Sun and the Earth align with Jupiter's nodal axis at the
beginning of January, and again at the beginning of June, when
the Sun is at 10° of Capricorn and Cancer. It would be reasonable
to expect those born on these dates to exhibit markedly the
characteristics of Jupiter, particularly its wisdom, optimism,
theatricality, and love of travel. A little research turns up Kipling,
whom we have considered previously with regard to Phaethon,
Georges Sand, and the pioneer aviator Amy Johnson, all of whom
demonstrate the point quite well.

For the rest of the year, of course, the nodes are not exactly opposite as seen from Earth, nor is the Sun in conjunction with them. Wherever it falls, or whatever planet it aspects, the north node of Jupiter seems to offer an opportunity for growth and increase similar to that offered by Venus, but with one important difference: Venus attracts influences from outside, but Jupiter expands them from inside. It gives an opportunity to amass more by growing bigger, by developing that which the native already has (and that which is real: the development of *potential* appears to be the domain of Neptune's nodes — see p. 110). To ignore Jupiter's node is to ignore the chance to develop fully and profitably one's latent talent.

The south node demands that the native grow into the situation at hand, to give his talent to it, and in doing so, to bring profit to both himself and his society. Precisely what it is that the native does to achieve this is shown, as before, by the sign and house position of the natal *planet* Jupiter.

Saturn's nodes

At this distance from the Earth the nodes of a planet only swing through about a dozen degrees in the course of a year. The nodes of Saturn, like those of Jupiter, have their axis on Cancer-Capricorn, and therefore their alignment with the Sun in January and July. In this case the alignment occurs on the 14th or 15th of the relevant month, with the Sun at 23° of the appropriate sign.

It is worth noting that since these larger planets have their nodes in the same sign the whole year round, then it must follow that events or opportunities for expression of the nature of those planets can only occur in those signs. This means that aspects to other planets in the nativity, or transits, or whatever, are tied to a specific pair of signs. The inference to be drawn from this must surely be that these planets are in some way similar to the signs of their nodes, and their energy can only be experienced in certain ways. It is not difficult to appreciate the connection between the nature of Saturn, as it is usually understood, and that of the Cancer-Capricorn axis, where it produces opportunities for experience through its nodes — but to appreciate the connection of Jupiter to the same axis makes the brow furrow a little. The key in this case is the exaltation of Jupiter in Cancer, something often forgotten, or not used when imagining which qualities the planet represents.

There is a rich new vein of symbolic reference here. Any

understanding of the planets and their astrological nature cannot but be enhanced by considering the way they present themselves through the placement of their nodes, and in the case of the outer planets, where that placement is restricted to a few degrees, such consideration is essential.

Saturn's north node will present an opportunity for the native to acquire, or grow, a Saturnine structure or set of rules. It enables him to define the limits and range of his actions, and to regulate them if they were previously disordered. The experience itself may involve submission to a firmer discipline than usual.

The south node demands that the individual impose some sort of order and constraint upon a disordered situation, to prevent its collapse. In this situation, circumstances are too disordered to offer any support to the individual, and he must restore harmony and regulation by his own efforts to prevent total loss.

Chiron's nodes

Chiron's eccentric orbit means that its nodes sweep through a greater arc than is usual for such a distant body. It also means that from our point of view the north node appears to move over a wider arc than the south node. The axis is actually at the end of Aries and Libra, and the two nodes are in opposition to each other, with one of them conjunct the Sun, when the Sun is at 29° of Aries or Libra, which corresponds to about the 20 April and the same date in October each year.

If Chiron represents, by virtue of its peculiar orbit, the route by which the individual elevates himself beyond Saturn's sphere to the Uranian, then Chiron's north node must provide an opportunity to do this. Correspondingly, the south node must be where the individual gives others a helping hand to assist them in their own struggle to ascend, and where the knowledge gained by elevation is handed on for future generations.

Uranus's nodes

The nodes of Uranus, like most of the outer planet nodes, stay within a few degrees. Throughout the year they stay between 10° and 16° of Gemini and Sagittarius, with perfect opposition occurring on about 4 June and 6 December, when the Sun is at 13° of those signs.

The fact that they are limited to a few degrees raises interesting questions about the nature of Uranus itself. It can only present opportunities for personal development, choices involving Uranian energies, through the middle of Gemini and Sagittarius, it seems. What is it about those two signs which so facilitates Uranus's expression? Is it that they are the most mutable forms of the lighter elements, and so can best translate the nature of a force for change? Or is it that only the two lighter elements could possibly be rarefied enough to escape weighty Saturn and take on the nature of that which is beyond the individual and the material? Possibly so.

There is something else worth noting, too: the nodes of Uranus are in the same signs as the traditional exaltation of the Moon's nodes.

In the usual practice of astrology, without planetary nodes, the nodes of the Moon are often assigned the task of pointing forwards and backwards in time beyond the native's current existence. Given that the seven traditional planets represent man and his mortal limitations, especially that of time, shown by Saturn, then is it not interesting that the nodal axis of the first planet beyond time points in the same direction as the exaltation axis of the indicators within time? Repetition of symbolism at different levels is astrology at its purest; it is difficult to ignore the importance of the message when it is as firmly stated as this.

Uranus's nodal axis is also very close to that of Venus, which prompts consideration of what the two planets might produce together. One possibility is music, with the harmony of the one allied to the rhythm of the other. Another possibility is the outbreak of war, as Venusian peace is broken into by Uranus's need to change the order of things. On 5 June and 6 December, the Sun is in alignment with both of Venus's nodes and both of Uranus's nodes at the same time, more or less, and at sunrise on such days the local ascendant could be made to join the pattern. The attack on Pearl Harbor, 6 December 1941, is probably worth studying in this regard.

Uranus's north node must offer opportunities to move into completely new areas, involving new technologies and new ideas. It is a chance to make 'giant leaps for mankind'. The man who first used that phrase, Neil Armstrong, the first man on the Moon, has Uranus's north node conjunct natal Mars in the first house. Symbolism as simple and as powerful as this is very reassuring when beginning work with planetary nodes!

Uranus's south node demands the impossible: flashes of genius are required to rescue and revive things. These are the occasions when the native produces work of which he did not believe himself capable, and when adverse circumstances bring out the best in him.

Neptune's nodes

Neptune's nodes move between 8° and 12° of the Leo-Aquarius axis, with exact opposition at 10° of the signs, corresponding to the end of January and the beginning of August.

It is an odd thing to find so nebulous a planet expressing itself through two fixed signs, neither of which are particularly associated with the personal sensitivity which is so often quoted as Neptune's own quality. Perhaps a closer look at Neptune is called for.

Neptune's north node gives an event where the future potential of a situation is offered. Potential is exactly that, undeveloped and unformed, and it will depend on whether the offer is taken up as to whether the potential is converted into something actual and concrete. Universal potential, symbolized by Aquarius, can become crystallized in the specific individual, symbolized by Leo; conversely, his potential is made real by his interaction with the society (Aquarius) of which he is the focus. Perhaps the best example of all is the simplest — children. They are the potential future of mankind (and, for reincarnationists, were once a part of its past), and they are particularly associated with the Leo-Aquarius axis.

The south node of Neptune demands acknowledgement of what the native has received or inherited from his past; some of his unused wealth (not necessarily money) must be surrendered because it is needed by others.

Pluto's nodes

Pluto is a very long way from the Sun indeed. Consequently, its nodes only vary by a degree or so from their axis at 19° of Cancer and Capricorn during the course of the year. The dates when they are in exact opposition, and when the Sun and the Earth cross that axis too, are around 10 January and 10 July.

Pluto's north node must denote an opportunity to learn and use an entirely new order of things, produced by the overthrow of the old. After such a course is adopted a return to the old order is not possible, since the nature of Pluto is to effect changes and transformations which are permanent. Lenin's horoscope shows Pluto's north node conjunct his natal Uranus, a clear indication of the leader of the revolution.

Pluto's south node must produce occasions when the completely radical approach is all that can produce any kind of progress in a situation which is foundering: on such occasions the individual must decide what can be saved and what cannot.

It is interesting once more to note how a planet so strong in its actions can only show itself, at least during this century, through a pair of cardinal signs. The power for redefinition, for rewriting fresh words over what has gone before, is inherent in all cardinal signs, and is what Pluto requires for its expression. The closeness between the zodiacal position of the nodes of Saturn, the last traditional planet, and Pluto, the final modern one, is also intriguing. Mundane astrologers may like to investigate events occurring on 10-14 January and July, when the nodes of Saturn and Pluto are at their closest alignment with the Sun. The Mars-Saturn conjunction which fell between the two north nodes in January 1946 seems to provide a useful starting date.

The Moon's nodes
The nodes of the Moon are much more complex in their motions than those of the other planets. Although the Moon orbits the Sun like everything else in the solar system, it orbits the Earth first, so to speak, and that makes its overall path rather intricate. The Moon's nodes are also the only nodes which can *truly* retrogress, as opposed to the apparent retrogression which the planetary nodes display, and this offers extra nuances of interpretation which are not applicable to the nodes of the planets.

The Moon's unique position, making it an intermediary between the Earth and the other planets, makes it an intermediary in a symbolic sense too. The Moon acts as a collector and amplifier, catching the music of the higher spheres and feeding it to our emotional and animal systems so that we can *feel* what is going on.

The Moon's nodes have always been considered as points of significance, and their rapid motion means that they aspect, if not actually transit, the whole zodiac in the space of a few years, giving every individual as many opportunities to develop in new directions as there are significant points in his horoscope. The sequence of events between eclipses, nodal alignments, and the subsequent transits of the planets to their nodal positions, is one which offers very powerful indications of the events which match them in the sublunary sphere.

The events offered by the Moon's nodes are not always in plain view, however; events provided by a direct lunar node come to meet the native, but those from a retrograde lunar node have to be actively sought. As there is a double orbital element in the path of the Moon's nodes, so there is a double choice in the events to which they correspond. As above, so below, as always.

11.

PLANETARY NODE EXAMPLES

King George VI

George VI did not expect to become King, nor had he been trained for the job. There seemed no reason why his older brother should not become King on the death of their father, George V. In the event, however, as is well known, the Prince of Wales became Edward VIII for only a few months, and then abdicated as a result of his love for Mrs Simpson.

The Duke of York found himself in an unusual situation: he was

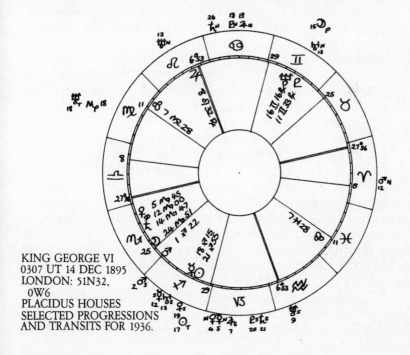

KING GEORGE VI
0307 UT 14 DEC 1895
LONDON: 51N32,
 0W6
PLACIDUS HOUSES
SELECTED PROGRESSIONS
AND TRANSITS FOR 1936.

in a position to become King. Such opportunities are rare, even in royal families — the heir to the throne is usually aware of his position since birth, and his siblings do not generally suppose that they will take his place in the normal run of things.

The horoscope of the Duke of York has had less attention paid to it than that of his brother in recent years, but it is just as interesting, if not more so. Although the Sun is not angular, its dispositor and the ruler of the ascendant are both so placed, and they are both benefic by nature, suggesting that he would lead a happy and fortunate life, in the main. As evidence of kingship, there is no need to look further than Jupiter (royal planet) in Leo (royal sign) conjunct the midheaven. The symbolism is strengthened and multiplied by the mutual reception of Jupiter and the Sun.

At the time of his brother's abdication, his progressed midheaven was square to his natal Mercury, suggesting that he might adopt a higher position at his brother's expense. His eighth house Neptune is busy, too: the progressed Moon passes it, the transiting Sun opposes it, while the planet itself, by transit, has reached his progressed midheaven, and thus the square of Mercury. His eighth house is his brother's sixth, of course, since his brother is his third house, and his brother's job, to use a sixth house word, is being king; Neptune dissolves him, and he is king no longer.

Whilst the secondary progressions just discussed are probably adequate indicators of the events, they seem in this instance to lack that clear and simple symbolism which characterizes the most potent astrology; perhaps the symbolic vocabulary selected was not that which most closely matches the events it is used to describe.

The use of Jupiter's nodes produces some striking patterns. They should, of course, provide indication of an opportunity to become king, if the usual symbolic values are adhered to, and they do. The lunar eclipse previous to the abdication, on 4 June, occurred with the Sun on the future king's Jupiter north node, at 12° Cancer. Here is a clear indication that within the time frame specified by the eclipse, i.e. the following few months, an opportunity would arise for him to use his tenth-house Jupiter to its fullest extent. Note how the solar and lunar eclipses are being used as the intermediary vehicle between the planetary nodes and the natal pattern. Three days after the abdication, on his 41st birthday, there is a solar eclipse on his natal Sun, which symbolizes him taking the solar role, that of being king. Two weeks later, on 28 December, there is a lunar eclipse with the Sun at 6° Capricorn, conjunct the south node of Jupiter in his natal horoscope. Here the nation demands that he give himself to their service as their king, and he accepts, using very similar words himself.

Whilst not everybody has the opportunity to become king, eclipses are a reliable indication of significant points in a life, and if they can be tied in to planetary nodes at the same time, then their message may be more clearly read.

President Johnson

Lyndon B. Johnson became President of the United States on the assassination of John F. Kennedy. The situation is not unlike that of George VI, in that the man who never expected to hold the highest office is presented with such an opportunity as a result of events in the life of another rather than his own. Again the pattern of solar and lunar eclipses and the planetary nodes is informative.

There were four eclipses in 1963: two in January and two in July. The first, a lunar eclipse at 18° Cancer, picks up his natal Venus, which is lady of his midheaven and so suggests some development in his career, or a change of title perhaps. This same eclipse is conjunct

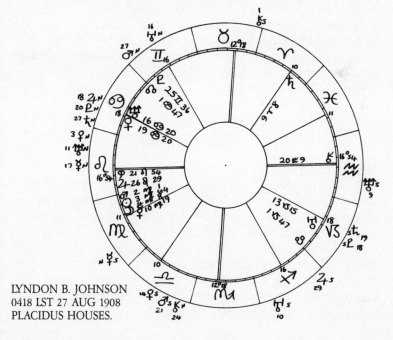

LYNDON B. JOHNSON
0418 LST 27 AUG 1908
PLACIDUS HOUSES.

his Jupiter north node, suggesting that he has an opportunity to expand his powers to their fullest extent and grow mightily in influence, but at the same time it is opposite his Pluto south node, which may be interpreted to mean that the event which offers him all this will exact a terrible payment as it occurs.

As if the first eclipse of Venus were not a strong enough hint, the solar eclipse on 25 January, at 4° Aquarius, opposes his Venus north node. A promise of wealth and fortune, provided that he feels able to respond to the challenge offered, as seems to be the case with opposition contacts to eclipses.

The lunar eclipse at 14° Capricorn on 6 July conjoins the natal Uranus. This planet is ruler of Johnson's seventh house. Here, then, is the sudden event in the life of his working partner (seventh house) — and with whom else does the Vice-President work if not the President himself?

The final eclipse, on 20 July, is at the same degree as the north node of Saturn, at 27° Cancer. Here is the opportunity to become the supreme administrator, but there are also clues to the likely effect on his working life and public duty (Saturn rules his sixth house), and the irreversible and profoundly memorable nature of the event itself (planet Saturn is tenant of his eighth).

As before, the solunar eclipses pick up the natal planetary nodes to translate the opportunities and demands they signify into events in time. The house position and rulerships of the planets whose nodes are activated will give important clues to the likely realization of the event.

Using the Planetary Nodes in Character Synthesis

Raymond Henry

Elsie Wheeler

Elsie Wheeler was the medium who worked with the astrologer Marc Edmund Jones to produce the Sabian Symbols, that powerful series of 360 images which provides all sorts of inexplicably appropriate detail when used with a horoscope.

Like so many of her profession, she has Neptune rising, and the nodes of the Moon in angular houses; it is by the house of the south node that we often find the quality and kind of mediumship to be expected. In this case it is in 'scientific' Aquarius and the authoritative tenth house, so that we expect Mrs Wheeler to go beyond the usual

comforting of the bereaved by proving survival after death; she will
have an interest in knowing how mediumship really works, what
the afterlife is really like, and what other useful purposes mediumship
can serve. The lunar south node will tap the 'higher learning' of
the ninth house for the inspiration it obliges the lady to pour out
to us. (With all mediums the lunar south node appears to tap the
house preceding the one it occupies.)

In Mrs Wheeler's case we also find that the nodes of her rising
Neptune have a major role to play. Neptune's south node is in the
tenth house, and even closer to the angle than that of the Moon.
Neptune's north node is in the fourth, preceded there by the north
node of Venus.

This lady projects a 'screen image' of herself to us on planet
Neptune in the first house, so that on meeting her we never really
see the Mrs Wheeler behind the role image. Strong Neptunes often
denote those film and TV actors whose faces are so familiar that we
are drawn to their latest film by their name rather than the merit
of the production itself — yet we never know the real person.
Mediums function in a very similar manner: we rarely know the

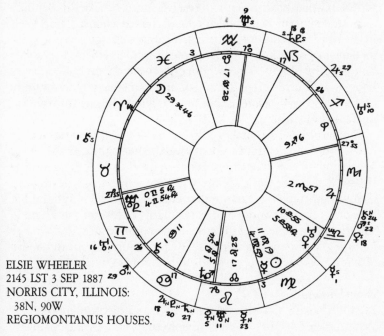

ELSIE WHEELER
2145 LST 3 SEP 1887
NORRIS CITY, ILLINOIS:
  38N, 90W
REGIOMONTANUS HOUSES.

medium as a person, nor do we need to, for we are in their presence
only to sense the presence of one who has been lost to physical reality
but whom we can enjoy as we enjoy the characters portrayed on the
cinema screen.

Neptune's north node here enables Mrs Wheeler to enter into
a strange fourth house — a material world — of the future (north
node): a world beyond the confines of our Saturn-limited lifespan,
a world where life goes on with the new vitality of the sign of Leo.
The Moon's north node here enables her to feel this state of affairs
quite physically; Venus's north node puts her at ease with it, and
Mercury's north node enables her to 'read the script there', so to speak.

And then the south nodes: Neptune's demands that she use her
remarkable gifts with authority and, tapping the heavenly ninth as
it does, assures us that what we hear from her is true. All that has
gone before, all *who* have gone before are thrusting through that
node to project to us through the person of the lady herself via her
rising planet.

The precise sextile between the north node of Mercury and the
south of Mars, conjunct with the north of Chiron, obliges her to
act upon what she discovers and receives, and to make this her life's
work (the Martian south node being in the sixth house). Mars itself
in the third and conjunct Saturn gives ample energy and authority
for her job of communication. Moreover, that Mars governs her twelfth
house — the Elsie Wheeler whom only she and God can ever truly
know — in which we find Chiron's south node, that very community
of those who went before us, now insisting that she give from within
to the world without.

Tenant of her sixth house, and lord of her eighth, is Jupiter. Its
north node is on her third cusp, its south on her eighth. Mrs Wheeler
has divine *carte blanche* to speak as she will to us, just so long as
she speaks of that eighth house with its acceptance of death, and
its Jovian optimistic assurance that even in death all is well, for life
goes on in another world. Note the polarity of Jupiter's south node
with the Martian north at the cusp of the house of substance. I never
had the pleasure of hearing this lady prove survival but the nodes
of her horoscope prove it as well as her words ever could.

Peter Hurkos

Psychic abilities of a Mercurial rather than a Neptunian kind are
shown by the relevant nodes in this horoscope. Hurkos, a Dutchman,

fell from a ladder during the Second World War and awoke in hospital
to find himself uncontrollably clairvoyant. Though nearly murdered
by the Dutch resistance after he had spontaneously betrayed several
of them in the ward, he learned to control his abilities and used
them in the service of the resistance until the end of the war in Europe.
He later moved to the USA where he became a professional clairvoyant
investigator.

   Mercury's north node is conjunct the Sun in the twelfth, while
its south node is conjunct Uranus's north node in Gemini in the
same house. Planet Uranus is in the eighth. There is a straightforward
pattern linking the man, his accident, and the opportunities for the
communication of secrets that it gave him. As with the previous
example, Neptune is prominent, and so are its nodes, but this time
they are square to the planets Mercury and Saturn (and with an
exactness which helps interpretation — Neptune north is square
Mercury, Neptune south square Saturn), which indicates his conscious
efforts to understand and control his ability.

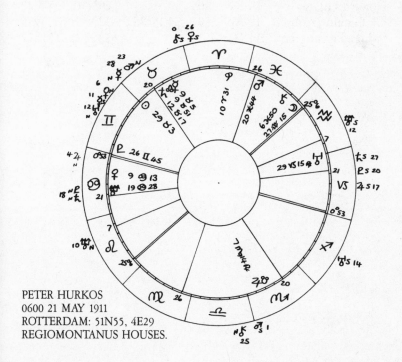

PETER HURKOS
0600 21 MAY 1911
ROTTERDAM: 51N55, 4E29
REGIOMONTANUS HOUSES.

Other nativities referred to with reference to planetary nodes

Below we give data for Houdini, who needs no introduction, and
Sidney Piddington, who was a popular 'mind-reader' some years
ago. It has never been clear whether Piddington was simply a clever
and practised entertainer, or whether he was to any extent genuinely
telepathic. Examination of his horoscope shows the two nodes of
Mercury very close indeed, with the north node conjunct his natal
Sun, a good enough indication of the field this Taurean chose to
occupy. No notice need be taken of his possible telepathy, were it
not for Chiron south on his planet Mercury in the twelfth . . .

Houdini, intriguing as ever, shows the opposite pattern, with north
Mercury on planet Chiron. It is worth remembering, in this regard,
his constant campaign against fraudulent mediums, as well as his
more famous feats.

Name	Date	Time	Place
Houdini	06 Apr 1874	02.24 LT	47N30, 19E05
Howard Hughes	24 Dec 1905		29N50, 95W20
Ivy G'stein-Jacobson	12 Apr 1893	11.15 PST	32N42, 117W15
Sidney Piddington	14 May 1918		33S52, 151E12

12.

A FINAL EXAMPLE: J. B. RHINE

Raymond Henry

Few names rank higher in the field of academic research into parapsychology than that of J. B. Rhine. His laboratory experiments made 'ESP' a familiar phrase world-wide from the time he published *New Frontiers of the Mind* in February 1938, and his specially devised 'Zener' cards for experiments in telepathy quickly attained mass sales rivalled only by the book itself. Telling friends what they were thinking became a new party trick for all to play; Rhine's methods, though now somewhat dated, are still in vogue both for fun and for serious research.

This man represents the start of parapsychology in many ways, and it seems reasonable to examine his sunrise horoscope with the help of all the Dark Stars this volume has discussed: Phaethon, the Dark Suns, and the planetary nodes.

It is widely believed that in some previous 'golden age' we were naturally telepathic. To consider such an age we must surely reinsert Phaethon into the heavens to give the balanced, and less frenetic, environment we knew then. One of Rhine's discoveries in telepathy was that it seemed to operate with less concern for time and sequence than we expect nowadays. Perhaps it comes from a part of our cerebral development which worked to a different system of time.

The energy source of Rhine's work, his Mars, is sandwiched between the Sun and Phaethon, which is on a degree of which Rudhyar writes 'Problems attending the transmission of knowledge in a special cultural set-up' (Rudhyar, *An Astrological Mandala*).

Where did Rhine direct his efforts? The Dark Sun of Mars is in the final minutes of Virgo, in the twelfth; he strives to define the technique (Virgo) by which the subconscious communicates. This is the field in which Rhine makes his return to the system for the abundant energy which the Sun-Mars conjunction bestows on him.

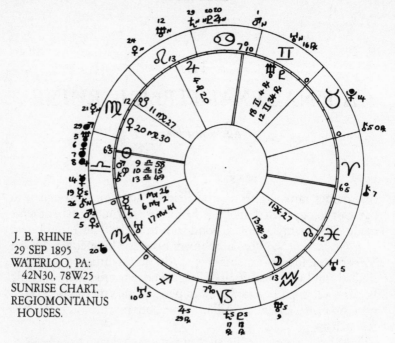

J. B. RHINE
29 SEP 1895
WATERLOO, PA:
42N30, 78W25
SUNRISE CHART,
REGIOMONTANUS
HOUSES.

The Dark Suns of Venus, the Earth, and Jupiter are all nestled between the Sun, Mars, and Phaethon: aptitude, inclination, and expansive uncluttered approach all call upon the service of his high energy. Neptune's Dark Sun is just behind the radiant Sun on yet another degree which Rudhyar associates with 'transmission' — this time of 'inner knowledge on which a New World could be built' (Rudhyar, op. cit.).

Chiron's Dark Sun is opposite the Earth's within nine minutes of arc, and on the seventh house cusp of this sunrise chart: Rhine's work bids fair to upgrade the whole state of human relationships, and move us towards that planetary consciousness which is now agreed, in principle at least, to be essential for the survival of mankind and the planet alike.

The planetary nodes are as informative as ever. North Mercury is conjunct his Venus in the twelfth, dispositor of his Sun. His energies find the opportunity for their expression in subconscious communication, therefore, a similar indication to his Mars Dark Sun.

The solar eclipse of December 1937, immediately preceding the publication of *New Frontiers of the Mind*, fell at 10° Sagittarius, conjunct Uranus's south node. Obviously something new was demanded of him, and he produced it. It was an original contribution, too; Uranus is the ruler of the fifth house of this horoscope. Are the wavy lines of Aquarius the inspiration for the Zener card symbol?

AFTERWORD

Much of what has been presented here is new and untried; though it is our hope that other astrologers find these Dark Stars useful and worthy of further investigation, we accept that many will reject them out of hand.

In certain quarters of the astrological fraternity, there has recently been something of a counter-Reformation, whereby the original seven planets are the only ones considered, and any doctrine not found in print before 1700 is derided. Whilst this is no bad thing in itself, leading as it does to a sense of ongoing history and tradition, as well as concentrating attention on the original symbolism of the seven and the twelve, it is prone to closed thinking, and that is something which does astrology no good in the long run.

Conversely, as these words are written, the newspapers contain reports of a planet beyond Pluto detected by deep space probes. In time perhaps this planet too, if it proves to be such, will be given symbolic values and inserted into the astrologer's framework. There are those who eagerly await new planets, and who concentrate their work on the trans-Saturnian planets entirely; whilst they are unlikely to be in agreement with the seven-planet reactionaries, there can be no denying that both groups practise astrology.

Astrology is a language, used to describe one thing in terms of another — the sky. Obviously the astrologer is the key element in the process, since he draws the parallels and makes the judgements, but his expression will be restricted if his vocabulary is limited. He needs as wide a vocabulary of symbols as possible. He does not have to use all of the words in his language all of the time, but what he has to say can be better put if he has a variety of ways in which to say it.

What we have tried to do here is to extend the vocabulary. None of our new words are out of place, or from roots in an incompatible language; they have grown, as such things do, from what was already there. Phaethon has the asteroids to prove its existence, while the Dark Suns and nodes are as much a part of the planets' orbits as their zodiacal longitude.

Whichever set of symbols the astrologer chooses to use, he will still be able to express himself and his judgements. Seven planets, ten, more; midpoints, harmonics, nodes. Some choose one set of values, some another. It is up to the astrologer to decide whether planetary nodes, say, will suit his purpose. Astrology is not made wrong by their omission, nor made right by their inclusion, but if they help to express what the astrologer has to say at the time, then let him use them.

We would tentatively suggest, and no more, that Phaethon would prove useful where interest lies in where a thing came from rather than where it is going; that the Dark Suns would prove useful when the question is not so much 'What have I got?' but 'What am I to do with it?'; and that the planetary nodes may offer illumination in the timing and prediction of major life events.

These are simply our preferences — our real hope is that you will incorporate these Dark Stars into your own astrological vocabulary. We are sure that you will find them as vivid and full of meaning as any you have ever used.

TABLES

I.

ZODIACAL LONGITUDES OF PHAETHON 1900-2003

At 0 hrs UT on the 1st of each month

1900 Stations: 9 Feb: 16° ♎ 20′; 17 May: 2° ♎ 46′					
JAN ♎	FEB ♎	MAR ♎	APR ♎	MAY ♎	JUN ♎
11°31′	16°5′	15°4′ R	9°3′	3°40′	3°22′D
JUL ♎	AUG ♎	SEP ♎	OCT ♏	NOV ♏	DEC ♐
8°5′	16°25′	27°0′	8°35′	21°26′	4°16′

1901 Stations: 25 May: 28° ♑ 19′; 2 Sept: 15° ♑ 36′					
JAN ♐	FEB ♑	MAR ♑	APR ♑	MAY ♑	JUN ♑
17°31′	0°12′	10°38′	20°13′	26°26′	28°15′ R
JUL ♑	AUG ♑	SEP ♑	OCT ♑	NOV ♑	DEC ♒
24°49′	18°43′	15°37′	17°54′D	24°49′	4°29′

1902 Stations: 10 Sept: 11° ♑ 25′; 17 Dec: 27° ♈ 17′

JAN ♒	FEB ♒	MAR ♓	APR ♓	MAY ♈	JUN ♈
16°20′	29°12′	11°10′	24°19′	6°34′	18°20′

JUL ♈	AUG ♉	SEP ♉	OCT ♉	NOV ♉	DEC ♈
28°22′	6°30′	10°51′	9°47′ **R**	3°37′	28°6′ D

1903 Stations: 20 Dec: 22° ♌ 45′

JAN ♈	FEB ♉	MAR ♉	APR ♉	MAY ♊	JUN ♊
28°7′ D	3°51′	12°12′	23°25′	5°18′	18°2′

JUL ♋	AUG ♋	SEP ♋	OCT ♌	NOV ♌	DEC ♌
0°38′	13°10′	25°22′	6°16′	15°41′	21°35′

1904 Stations: 24 Mar: 8° ♌ 33′

JAN ♌	FEB ♌	MAR ♌	APR ♌	MAY ♌	JUN ♌
22°13′ **R**	16°42′	10°27′	8°44′ D	12°41′	20°41′

JUL ♍	AUG ♍	SEP ♍	OCT ♎	NOV ♎	DEC ♏
0°39′	12°19′	24°50′	7°23′	20°29′	2°54′

1905 Stations: 31 Mar: 4° ♐ 3′; 8 Jul: 21° ♏ 7′

JAN ♏	FEB ♏	MAR ♐	APR ♐	MAY ♐	JUN ♏
14°50′	24°57′	1°26′	4°3′ **R**	1°5′	24°51′

JUL ♏	AUG ♏	SEP ♏	OCT ♐	NOV ♐	DEC ♑
21°14′	22°51′ D	29°7′	8°10′	19°31′	1°41′

1906 Stations: 16 Jul: 16° ♓45′; 21 Oct: 3° ♓40′

JAN ♑	FEB ♑	MAR ♒	APR ♒	MAY ♓	JUN ♓
14°55′	28°16′	10°1′	22°12′	2°36′	11°5′

JUL ♓	AUG ♓	SEP ♓	OCT ♓	NOV ♓	DEC ♓
15°57′	15°59′ **R**	10°52′	5°8′	3°50′D	7°58′

1907 Stations: 31 Oct: 29° ♓2′

JAN ♓	FEB ♓	MAR ♈	APR ♈	MAY ♉	JUN ♉
16°34′	27°35′	8°43′	21°38′	4°15′	17°3′

JUL ♉	AUG ♊	SEP ♊	OCT ♊	NOV ♊	DEC ♊
28°56′	10°16′	19°59′	26°43′	29°1′ **R**	25°15′

1908 Stations: 1 Feb: 14° ♊47′

JAN ♊	FEB ♊	MAR ♊	APR ♊	MAY ♋	JUN ♋
18°6′ **R**	14°47′D	17°28′	24°54′	4°41′	16°16′

JUL ♋	AUG ♌	SEP ♌	OCT ♍	NOV ♍	DEC ♍
28°15′	11°1′	23°54′	6°11′	18°15′	28°35′

1909 Stations: 6 Feb: 10° ♎13′; 13 May: 26° ♍36′

JAN ♎	FEB ♎	MAR ♎	APR ♎	MAY ♍	JUN ♍
6°40′	10°12′	8°6′ **R**	1°34′	26°58′	27°48′D

JUL ♎	AUG ♎	SEP ♎	OCT ♏	NOV ♏	DEC ♐
3°20′	12°13′	23°8′	4°54′	17°50′	0°41′

1910 Stations: 19 May: 22° ♑ 8'; 28 Aug: 8° ♑ 37'

JAN	FEB	MAR	APR	MAY	JUN
♐	♐	♑	♑	♑	♑
13°52'	26°23'	6°32'	15°37'	21°3'	21°44' R
JUL	AUG	SEP	OCT	NOV	DEC
♑	♑	♑	♑	♑	♒
17°24'	11°29'	9°29' D	12°47'	20°26'	0°30'

1911 Stations: 4 Sep: 4° ♉ 55'; 10 Dec: 21° ♈ 12'

JAN	FEB	MAR	APR	MAY	JUN
♒	♒	♓	♓	♈	♈
12°36'	25°35'	7°35'	20°41'	2°50'	14°3'
JUL	AUG	SEP	OCT	NOV	DEC
♈	♉	♉	♉	♈	♈
24°3'	1°32'	4°56'	2°45' R	26°9'	21°30'

1912 Stations: 15 Dec: 16° ♌ 39'

JAN	FEB	MAR	APR	MAY	JUN
♈	♈	♉	♉	♊	♊
22°44' D	29°18'	8°28'	20°1'	2°3'	14°52'
JUL	AUG	SEP	OCT	NOV	DEC
♊	♋	♋	♌	♌	♌
27°18'	9°54'	21°55'	2°28'	11°17'	16°11'

1913 Stations: 19 Mar: 2° ♌ 6'

JAN	FEB	MAR	APR	MAY	JUN
♌	♌	♌	♌	♌	♌
15°23' R	8°58'	3°31'	3°0' D	7°52'	16°26'
JUL	AUG	SEP	OCT	NOV	DEC
♌	♍	♍	♎	♎	♎
26°45'	8°37'	21°13'	3°48'	16°52'	29°9'

1914 Stations: 26 Mar: 27° ♏55'; 1 Jul: 14° ♏55'

JAN ♏	FEB ♏	MAR ♏	APR ♏	MAY ♏	JUN ♏
10°49'	20°28'	26°14'	27°44' R	23°45'	17°32'

JUL ♏	AUG ♏	SEP ♏	OCT ♐	NOV ♐	DEC ♐
14°55'D	17°36'	24°37'	4°7'	15°44'	28°3'

1915 Stations: 9 Jul: 10° ♓29'; 17 Oct: 27° ♒27'

JAN ♑	FEB ♑	MAR ♒	APR ♒	MAY ♒	JUN ♓
11°20'	24°39'	6°19'	18°17'	28°20'	6°15'

JUL ♓	AUG ♓	SEP ♓	OCT ♒	NOV ♒	DEC ♓
10°13'	9°6' R	3°22'	28°16'	28°9'D	3°19'

1916 Stations: 22 Oct: 22° ♊57'

JAN ♓	FEB ♓	MAR ♈	APR ♈	MAY ♉	JUN ♉
12°24'	23°45'	5°28'	18°28'	1°5'	13°49'

JUL ♉	AUG ♊	SEP ♊	OCT ♊	NOV ♊	DEC ♊
25°31'	6°33'	15°44'	21°35'	22°33' R	17°38' R

1917 Stations: 25 Jan: 8° ♊41'

JAN ♊	FEB ♊	MAR ♊	APR ♊	MAY ♋	JUN ♋
10°45' R	8°51'D	12°29'	20°34'	0°44'	12°33'

JUL ♋	AUG ♌	SEP ♌	OCT ♍	NOV ♍	DEC ♍
24°38'	7°26'	20°17'	2°28'	14°18'	24°15'

1918 Stations: 1 Feb: 4° ♎ 7′; 6 May: 20° ♍ 25′

JAN ♎	FEB ♎	MAR ♎	APR ♍	MAY ♍	JUN ♍
1°39′	4°7′ **R**	0°56′	24°10′	20°29′	22°27′ D
JUL ♍	AUG ♎	SEP ♎	OCT ♏	NOV ♏	DEC ♏
28°44′	8°7′	19°18′	1°14′	14°14′	27°5′

1919 Stations: 12 May: 15° ♑ 57′; 21 Aug: 3° ♑ 15′

JAN ♐	FEB ♐	MAR ♑	APR ♑	MAY ♑	JUN ♑
10°11′	22°31′	2°22′	10°52′	15°27′	14°59′ **R**
JUL ♑	AUG ♑	SEP ♑	OCT ♑	NOV ♑	DEC ♑
9°55′	4°27′	3°35′ D	7°51′	16°8′	26°35′

1920 Stations: 27 Aug: 28° ♈ 47′; 2 Dec: 15° ♈ 7′

JAN ♒	FEB ♒	MAR ♓	APR ♓	MAY ♓	JUN ♈
8°54′	21°59′	4°25′	17°28′	29°29′	10°42′
JUL ♈	AUG ♈	SEP ♈	OCT ♈	NOV ♈	DEC ♈
19°54′	26°34′	28°45′ **R**	25°20′	18°35′	15°7′ **R**

1921 Stations: 9 Dec: 10° ♌ 34′

JAN ♈	FEB ♈	MAR ♉	APR ♉	MAY ♉	JUN ♊
17°42′ D	25°10′	4°27′	16°15′	28°26′	11°17′
JUL ♊	AUG ♋	SEP ♋	OCT ♋	NOV ♌	DEC ♌
23°42′	6°12′	18°12′	28°15′	6°29′	10°26′

1922 Stations: 13 Mar: 26°♋18′

JAN ♌	FEB ♌	MAR ♋	APR ♋	MAY ♌	JUN ♌
8°25′ **R**	1°30′	26°47′	27°29′ **D**	3°11′	12°17′

JUL ♌	AUG ♍	SEP ♍	OCT ♎	NOV ♎	DEC ♎
22°53′	4°56′	17°38′	0°13′	13°1′	25°21′

1923 Stations: 19 Mar: 21° ♏47′; 25 Jun: 8° ♏43′

JAN ♏	FEB ♏	MAR ♏	APR ♏	MAY ♏	JUN ♏
6°43′	15°51′	20°51′	21°11′ **R**	16°9′	10°22′

JUL ♏	AUG ♏	SEP ♏	OCT ♐	NOV ♐	DEC ♐
8°49′ **D**	12°32′	20°13′	0°7′	11°59′	24°25′

1924 Stations: 1 Jul: 4° ♓16′; 10 Oct: 21° ♒9′

JAN ♑	FEB ♑	MAR ♒	APR ♒	MAY ♒	JUN ♓
7°44′	21°2′	2°59′	14°40′	24°16′	1°24′

JUL ♓	AUG ♓	SEP ♒	OCT ♒	NOV ♒	DEC ♒
4°16′ **R**	1°52′	25°42′	21°33′	22°50′ **D**	28°57′

1925 Stations: 16 Oct: 16° ♊49′

JAN ♓	FEB ♓	MAR ♈	APR ♈	MAY ♈	JUN ♉
8°39′	20°21′	1°49′	14°53′	27°29′	10°8′

JUL ♉	AUG ♊	SEP ♊	OCT ♊	NOV ♊	DEC ♊
21°40′	2°24′	11°3′	16°6′	15°54′ **R**	10°10′ **D**

1926 Stations: 19 Jan: 2° ♓ 46'

JAN ♊	FEB ♊	MAR ♊	APR ♊	MAY ♊	JUN ♋
3°49' R	3°8' D	7°39'	16°20'	26°51'	8°51'

JUL ♋	AUG ♌	SEP ♌	OCT ♌	NOV ♍	DEC ♍
21°1'	3°51'	16°4'	28°43'	10°18'	19°50'

1927 Stations: 26 Jan: 28° ♍ 2'; 30 Apr: 14° ♍ 14'

JAN ♍	FEB ♍	MAR ♍	APR ♍	MAY ♍	JUN ♍
26°29'	27°47' R	23°37'	16°53'	14°15' D	17°16'

JUL ♍	AUG ♎	SEP ♎	OCT ♎	NOV ♏	DEC ♏
24°15'	4°4'	15°31'	27°35'	10°40'	23°29'

1928 Stations: 6 May: 9° ♑ 47'; 14 Aug: 27° ♐ 4'

JAN ♐	FEB ♐	MAR ♐	APR ♑	MAY ♑	JUN ♑
6°29'	18°36'	28°24'	6°12'	9°41'	7°53' R

JUL ♑	AUG ♐	SEP ♐	OCT ♑	NOV ♑	DEC ♑
2°13'	27°33'	28°1' D	3°19'	12°16'	23°6'

1929 Stations: 21 Aug: 22° ♈ 27'; 27 Nov: 9° ♈ 10'

JAN ♒	FEB ♒	MAR ♓	APR ♓	MAY ♓	JUN ♈
5°38'	18°49'	0°50'	13°48'	25°38'	6°36'

JUL ♈	AUG ♈	SEP ♈	OCT ♈	NOV ♈	DEC ♈
15°20'	21°15'	22°19 R	17°56'	11°23'	9°3' D

1930 Stations: 3 Dec: 4° ♌ 29'

JAN ♈	FEB ♈	MAR ♉	APR ♉	MAY ♉	JUN ♊
12°42'	20°51'	0°30'	12°32'	24°49'	7°43'
JUL ♊	AUG ♋	SEP ♋	OCT ♋	NOV ♌	DEC ♌
20°5'	2°29'	14°5'	23°57'	1°31'	4°28'

1931 Stations: 7 Mar: 20° ♋ 11'

JAN ♌	FEB ♋	MAR ♋	APR ♋	MAY ♋	JUN ♌
1°16' R	24°7'	20°19'	22°9' D	28°37'	8°12'
JUL ♌	AUG ♍	SEP ♍	OCT ♍	NOV ♎	DEC ♎
19°4'	1°16'	14°1'	26°38'	9°34'	21°30'

1932 Stations: 13 Mar: 15° ♏ 40'; 18 Jun: 2° ♏ 31'

JAN ♏	FEB ♏	MAR ♏	APR ♏	MAY ♏	JUN ♏
2°33'	11°0'	15°20'	14°17' R	8°36'	3°19'
JUL ♏	AUG ♏	SEP ♏	OCT ♏	NOV ♐	DEC ♐
3°4' D	7°52'	16°14'	26°34'	8°40'	21°1'

1933 Stations: 25 Jun: 28° ♒ 4'; 4 Oct: 15° ♒ 11'

JAN ♑	FEB ♑	MAR ♑	APR ♒	MAY ♒	JUN ♒
4°35'	17°49'	29°12'	10°37'	19°47'	26°11'
JUL ♒	AUG ♒	SEP ♒	OCT ♒	NOV ♒	DEC ♒
28°2' R	24°34'	18°21'	15°12'	17°35' D	24°28'

1934 Stations: 10 Oct: 10° ♓ 40'

JAN ♓	FEB ♓	MAR ♓	APR ♈	MAY ♈	JUN ♉
4°38'	16°35'	28°11'	11°17'	23°53'	6°26'
JUL ♉	AUG ♉	SEP ♓	OCT ♓	NOV ♓	DEC ♓
17°47'	28°10'	6°16'	10°24'	9°0' **R**	2°40'

1935 Stations: 13 Jan: 26° ♉ 31'

JAN ♉	FEB ♉	MAR ♊	APR ♊	MAY ♊	JUN ♋
27°7' **R**	27°38' D	2°57'	12°11'	23°0'	5°10'
JUL ♋	AUG ♌	SEP ♌	OCT ♌	NOV ♍	DEC ♍
17°26'	0°16'	13°1'	24°56'	6°15'	15°18'

1936 Stations: 20 Jan: 21° ♍ 57'; 23 Apr: 8° ♍ 5'

JAN ♍	FEB ♍	MAR ♍	APR ♍	MAY ♍	JUN ♍
21°6'	21°12' **R**	15°57'	9°36'	8°18' D	12°28'
JUL ♍	AUG ♎	SEP ♎	OCT ♎	NOV ♏	DEC ♏
20°11'	0°27'	12°10'	24°24'	7°30'	20°18'

1937 Stations: 30 Apr: 3° ♑ 37'; 9 Aug: 20° ♐ 53'

JAN ♐	FEB ♐	MAR ♐	APR ♑	MAY ♑	JUN ♑
3°9'	14°59'	24°2'	1°7'	3°36' **R**	0°41'
JUL ♐	AUG ♐	SEP ♐	OCT ♐	NOV ♑	DEC ♑
24°46'	21°0'	22°37' D	28°41'	8°9'	19°17'

1938 Stations: 15 Aug: 16° ♈ 25′; 22 Nov: 2° ♈ 55′

JAN ♒	FEB ♒	MAR ♒	APR ♓	MAY ♓	JUN ♈
1°59′	15°14′	27°14′	10°7′	21°46′	2°25′
JUL ♈	AUG ♈	SEP ♈	OCT ♈	NOV ♈	DEC ♈
10°39′	15°43′	15°39′ R	10°28′	4°21′	3°13′ D

1939 Stations: 28 Nov: 28° ♋ 22′

JAN ♈	FEB ♈	MAR ♈	APR ♉	MAY ♉	JUN ♊
7°5′	16°38′	26°38′	8°50′	21°13′	4°8′
JUL ♊	AUG ♊	SEP ♋	OCT ♋	NOV ♋	DEC ♋
16°28′	28°44′	10°6′	19°34′	26°25′	28°18′ R

1940 Stations: 1 Mar: 14° ♋ 5′

JAN ♋	FEB ♋	MAR ♋	APR ♋	MAY ♋	JUN ♌
23°57′ R	16°50′	14°5′ D	17°11′	24°28′	4°32′
JUL ♌	AUG ♌	SEP ♍	OCT ♍	NOV ♎	DEC ♎
15°41′	28°3′	10°53′	23°27′	6°16′	17°59′

1941 Stations: 6 Mar: 9° ♏ 34′; 12 Jun: 26° ♎ 19′

JAN ♎	FEB ♏	MAR ♏	APR ♏	MAY ♏	JUN ♎
28°36′	6°24′	9°30′	7°14′ R	1°7′	26°37′
JUL ♎	AUG ♏	SEP ♏	OCT ♏	NOV ♐	DEC ♐
27°28′ D	3°6′	12°3′	22°41′	4°59′	17°38′

1942 Stations: 19 Jun: 21°≈52'; 27 Sept: 9°≈2'

JAN ♑	FEB ♑	MAR ♑	APR ≈	MAY ≈	JUN ≈
1°0'	14°9'	25°23'	6°30'	15°11'	20°47'

JUL ≈	AUG ≈	SEP ≈	OCT ≈	NOV ≈	DEC ≈
21°33' R	17°9'	11°8'	9°4' D	12°32'	20°6'

1943 Stations: 4 Oct: 4° ♓ 32'

JAN ♓	FEB ♓	MAR ♓	APR ♈	MAY ♈	JUN ♉
0°40'	12°52'	24°33'	7°43'	20°16'	2°42'

JUL ♉	AUG ♉	SEP ♊	OCT ♊	NOV ♊	DEC ♉
13°50'	23°50'	1°19'	4°31'	1°55' R	25°13' R

1944 Stations: 1 Jan: 20° ♉ 37'

JAN ♉	FEB ♉	MAR ♉	APR ♊	MAY ♊	JUN ♋
20°37' R	22°20' D	28°40'	8°28'	19°35'	1°56'

JUL ♋	AUG ♋	SEP ♌	OCT ♌	NOV ♍	DEC ♍
14°16'	27°6'	9°45'	21°29'	2°26'	10°51'

1945 Stations: 12 Jan: 15° ♍ 53'; 18 Apr: 1° ♍ 56'

JAN ♍	FEB ♍	MAR ♍	APR ♍	MAY ♍	JUN ♍
15°36'	14°16' R	8°28'	2°45'	2°34' D	7°40'

JUL ♍	AUG ♍	SEP ♎	OCT ♎	NOV ♏	DEC ♏
15°55'	26°32'	8°28'	20°46'	3°55'	16°40'

1946 Stations: 24 Apr: 27° ♐ 27'; 3 Aug: 14° ♐ 42'

JAN ♏	FEB ♐	MAR ♐	APR ♐	MAY ♐	JUN ♐
29°22'	10°55'	19°33'	25°53'	27°18' **R**	23°21'

JUL ♐	AUG ♐	SEP ♐	OCT ♐	NOV ♑	DEC ♑
17°26'	14°40'	17°22' D	24°10'	4°6'	15°30'

1947 Stations: 9 Aug: 10° ♈ 14'; 15 Nov: 26° ♓ 50'

JAN ♑	FEB ♒	MAR ♒	APR ♓	MAY ♓	JUN ♓
28°20'	11°39'	23°38'	6°25'	17°51'	28°9'

JUL ♈	AUG ♈	SEP ♈	OCT ♈	NOV ♓	DEC ♓
5°49'	9°59'	8°46' **R**	2°59'	27°33'	27°36' D

1948 Stations: 21 Nov: 22° ♋ 17'

JAN ♈	FEB ♈	MAR ♈	APR ♉	MAY ♉	JUN ♊
3°10'	12°29'	23°8'	5°34'	18°2'	0°57'

JUL ♊	AUG ♊	SEP ♋	OCT ♋	NOV ♋	DEC ♋
13°13'	25°19'	6°22'	15°19'	21°14'	21°47' **R**

1949 Stations: 22 Feb: 7° ♋ 58'

JAN ♋	FEB ♋	MAR ♋	APR ♋	MAY ♋	JUN ♌
16°17' **R**	9°36'	8°7' D	12°14'	20°8'	0°35'

JUL ♌	AUG ♌	SEP ♍	OCT ♍	NOV ♎	DEC ♎
11°57'	24°26'	7°18'	19°50'	2°32'	14°2'

1950 Stations: 28 Feb: 3° ♏27′; 6 Jun: 20° ♎7′

JAN ♎	FEB ♏	MAR ♏	APR ♏	MAY ♎	JUN ♎
24°14′	1°19′	3°26′ R	0°1′	23°42′	20°10′
JUL ♎	AUG ♎	SEP ♏	OCT ♏	NOV ♐	DEC ♐
22°4′ D	28°31′	7°56′	18°52′	1°20′	14°3′

1951 Stations: 13 Jun: 15° ♒41′; 21 Sept: 2° ♒53′

JAN ♐	FEB ♑	MAR ♑	APR ♒	MAY ♒	JUN ♒
27°24′	10°27′	21°31′	2°19′	10°26′	15°11′
JUL ♒	AUG ♒	SEP ♒	OCT ♒	NOV ♒	DEC ♒
14°50′ R	9°40′	4°7′	3°11′ D	7°39′	15°49′

1952 Stations: 27 Sept: 28° ♉24′

JAN ♒	FEB ♓	MAR ♓	APR ♈	MAY ♈	JUN ♈
26°46′	9°10′	21°23′	4°33′	17°2′	29°18′
JUL ♉	AUG ♉	SEP ♉	OCT ♉	NOV ♉	DEC ♉
10°10′	19°41′	26°21′	28°20′ R	24°26′	17°40′ R

1953 Stations: 2 Jan: 14° ♉21′

JAN ♉	FEB ♉	MAR ♉	APR ♊	MAY ♊	JUN ♊
14°21′ R	17°23′ D	24°14′	4°28′	15°50′	28°18′
JUL ♋	AUG ♋	SEP ♌	OCT ♌	NOV ♌	DEC ♍
10°41′	23°30′	6°4′	17°36′	28°12′	6°3′

1954 Stations: 7 Jan: 9° ♍ 47′; 11 Apr: 25° ♌ 46′

JAN	FEB	MAR	APR	MAY	JUN
♍	♍	♍	♌	♌	♍
9°46′	7°13′ **R**	0°59′	26°5′	27°2′D	2°59′
JUL	AUG	SEP	OCT	NOV	DEC
♍	♍	♎	♎	♏	♏
11°45′	22°41′	4°47′	17°11′	0°19′	13°1′

1955 Stations: 18 Apr: 21° ♐ 16′; 27 Jul: 8° ♐ 29′

JAN	FEB	MAR	APR	MAY	JUN
♏	♐	♐	♐	♐	♐
25°33′	6°48′	14°57′	20°27′	20°44′ **R**	15°53′
JUL	AUG	SEP	OCT	NOV	DEC
♐	♐	♐	♐	♑	♑
10°15′	8°36′D	12°18′	19°48′	0°8′	11°46′

1956 Stations: 3 Aug: 4° ♈ 3′; 9 Nov: 20° ♓ 43′

JAN	FEB	MAR	APR	MAY	JUN
♑	♒	♒	♓	♓	♓
24°43′	8°4′	20°26′	3°4′	14°14′	24°5′
JUL	AUG	SEP	OCT	NOV	DEC
♈	♈	♈	♓	♓	♓
1°1′	4°2′	1°30′ **R**	25°18′	20°54′	22°18′D

1957 Stations: 14 Nov: 16° ♋ 11′

JAN	FEB	MAR	APR	MAY	JUN
♓	♈	♈	♉	♉	♉
28°52′	8°46′	19°20′	1°56′	14°27′	27°22′
JUL	AUG	SEP	OCT	NOV	DEC
♊	♊	♋	♋	♋	♋
9°32′	21°28′	2°12′	10°38′	15°41′	15°5′ **R**

1958 Stations: 17 Feb: 1°♋52′

JAN ♋	FEB ♋	MAR ♋	APR ♋	MAY ♋	JUN ♋
8°48′ R	2°45′	2°22′ D	7°28′	15°55′	26°42′

JUL ♌	AUG ♌	SEP ♍	OCT ♍	NOV ♍	DEC ♎
8°15′	20°49′	3°43′	16°13′	28°46′	10°2′

1959 Stations: 24 Feb: 27°♎19′; 30 May: 13°♎55′

JAN ♎	FEB ♎	MAR ♎	APR ♎	MAY ♎	JUN ♎
19°46′	26°2′	27°8′ R	22°40′	16°25′	13°56′ D

JUL ♎	AUG ♎	SEP ♏	OCT ♏	NOV ♏	DEC ♐
16°53′	24°3′	3°54′	15°5′	27°41′	10°28′

1960 Stations: 6 Jun: 9°♒29′; 14 Sept: 26°♑43′

JAN ♐	FEB ♑	MAR ♑	APR ♑	MAY ♒	JUN ♒
23°48′	6°45′	17°59′	28°20′	5°46′	9°23′

JUL ♒	AUG ♒	SEP ♑	OCT ♑	NOV ♒	DEC ♒
7°46′ R	1°58′	27°14′	27°36′ D	3°8′	11°58′

1961 Stations: 20 Sept: 22°♉14′; 26 Dec: 8°♉16′

JAN ♒	FEB ♓	MAR ♓	APR ♈	MAY ♈	JUN ♈
23°19′	5°55′	17°47′	0°58′	13°23′	25°29′

JUL ♉	AUG ♉	SEP ♉	OCT ♉	NOV ♉	DEC ♉
6°4′	15°7′	21°0′	21°55′ R	17°2′	10°30′ R

1962 Stations: None

JAN ♉	FEB ♉	MAR ♉	APR ♊	MAY ♊	JUN ♊
8°23'D	12°27'	19°54'	0°32'	12°6'	24°41'

JUL ♋	AUG ♋	SEP ♌	OCT ♌	NOV ♌	DEC ♍
7°6'	19°53'	2°19'	13°39'	23°52'	1°3'

1963 Stations: 2 Jan: 3° ♍42'; 5 Apr: 19° ♌38'

JAN ♍	FEB ♌	MAR ♌	APR ♌	MAY ♌	JUN ♌
3°43'	29°59'R	23°35'	19°40'	21°43'D	28°26'

JUL ♍	AUG ♍	SEP ♎	OCT ♎	NOV ♎	DEC ♏
7°39'	18°52'	1°8'	13°35'	26°44'	9°20'

1964 Stations: 11 Apr: 15° ♐8'; 19 Jul: 2° ♐17'

JAN ♏	FEB ♐	MAR ♐	APR ♐	MAY ♐	JUN ♐
21°42'	2°35'	10°27'	14°51'	13°50'R	8°11'

JUL ♐	AUG ♐	SEP ♐	OCT ♐	NOV ♐	DEC ♑
3°10'	2°50'D	7°39'	15°48'	26°35'	8°27'

1965 Stations: 25 Jul: 27° ♓50'; 2 Nov: 14° ♓37'

JAN ♑	FEB ♒	MAR ♒	APR ♒	MAY ♓	JUN ♓
21°33'	4°54'	16°47'	29°16'	10°12'	19°35'

JUL ♓	AUG ♓	SEP ♓	OCT ♓	NOV ♓	DEC ♓
25°49'	27°48'R	24°12'	17°56'	14°37'	17°7'D

1966 Stations: 8 Nov: 10°♋5′

JAN ♓	FEB ♈	MAR ♈	APR ♈	MAY ♉	JUN ♉
24°26′	4°47′	15°34′	23°18′	10°52′	23°46′
JUL ♊	AUG ♊	SEP ♊	OCT ♋	NOV ♋	DEC ♋
5°50′	17°34′	28°0′	5°50′	9°57′	8°9′ **R**

1967 Stations: 11 Feb: 25° ♓46′

JAN ♋	FEB ♊	MAR ♊	APR ♋	MAY ♋	JUN ♋
1°19′ **R**	26°8′	26°51′D	2°48′	11°46′	22°51′
JUL ♌	AUG ♌	SEP ♍	OCT ♍	NOV ♍	DEC ♎
4°34′	17°14′	0°8′	12°34′	24°59′	5°57′

1968 Stations: 17 Feb: 21° ♎14′; 23 May: 7° ♎44′

JAN ♎	FEB ♎	MAR ♎	APR ♎	MAY ♎	JUN ♎
15°10′	20°35′	20°29′ **R**	14°59′	9°10′	8°1′D
JUL ♎	AUG ♎	SEP ♏	OCT ♏	NOV ♏	DEC ♐
12°3′	19°58′	0°17′	11°44′	24°30′	7°18′

1969 Stations: 30 May: 3° ♒17′; 8 Sept: 20° ♑33′

JAN ♐	FEB ♑	MAR ♑	APR ♑	MAY ♒	JUN ♒
20°36′	3°23′	14°0′	23°56′	0°42′	3°17′ **R**
JUL ♒	AUG ♑	SEP ♑	OCT ♑	NOV ♑	DEC ♒
0°36′	24°32′	20°40′	22°11′D	28°32′	7°52′

1970 Stations: 14 Sept: 16° ♑ 5'; 20 Dec: 2° ♑ 11'

JAN ♒	FEB ♓	MAR ♓	APR ♓	MAY ♈	JUN ♈
19°31'	2°16'	14°12'	27°22'	9°42'	21°37'

JUL ♉	AUG ♉	SEP ♉	OCT ♉	NOV ♉	DEC ♉
1°54'	10°25'	15°28'	15°15' **R**	9°33'	3°31' **R**

1971 Stations: 26 Dec: 27° ♌ 38'

JAN ♉	FEB ♉	MAR ♉	APR ♉	MAY ♊	JUN ♊
2°37'D	7°41'	15°40'	26°39'	8°25'	21°6'

JUL ♋	AUG ♋	SEP ♋	OCT ♌	NOV ♌	DEC ♌
3°31'	16°15'	28°34'	9°39'	19°26'	25°54'

1972 Stations: 31 Mar: 13° ♌ 30'

JAN ♌	FEB ♌	MAR ♌	APR ♌	MAY ♌	JUN ♌
27°25' **R**	22°37'	16°6'	13°31'D	16°45'	24°17'

JUL ♍	AUG ♍	SEP ♍	OCT ♎	NOV ♎	DEC ♏
3°59'	15°29'	27°54'	10°6'	23°33'	6°3'

1973 Stations: 4 Apr: 8° ♐ 59'; 13 Jul: 26° ♏ 5'

JAN ♏	FEB ♏	MAR ♐	APR ♐	MAY ♐	JUN ♐
18°9'	28°36'	5°23'	8°57'	6°47' **R**	0°42'

JUL ♏	AUG ♏	SEP ♐	OCT ♐	NOV ♐	DEC ♑
26°26'	27°14'D	2°55'	11°37'	22°45'	4°47'

1974 Stations: 19 Jul: 21° ♓ 38'; 27 Oct: 8° ♓ 30'

JAN ♑	FEB ♒	MAR ♒	APR ♒	MAY ♓	JUN ♓
17°57'	1°19'	13°8'	25°27'	6°5'	14°59'

JUL ♓	AUG ♓	SEP ♓	OCT ♓	NOV ♓	DEC ♓
20°26'	21°19' **R**	16°47'	10°43'	8°33' **D**	12°5'

1975 Stations: 3 Nov: 3° ♋ 58'

JAN ♓	FEB ♈	MAR ♈	APR ♈	MAY ♉	JUN ♉
20°6'	0°51'	11°51'	24°41'	7°18'	20°9'

JUL ♊	AUG ♊	SEP ♊	OCT ♋	NOV ♋	DEC ♋
2°7'	13°38'	23°39'	0°52'	3°58'	1°2' **R**

1976 Stations: 4 Feb: 19° ♊ 41'

JAN ♊	FEB ♊	MAR ♊	APR ♊	MAY ♋	JUN ♋
23°55' **R**	19°45'	21°40' **D**	28°33'	8°2'	19°27'

JUL ♌	AUG ♌	SEP ♌	OCT ♍	NOV ♍	DEC ♎
1°20'	14°4'	26°57'	9°18'	21°31'	2°9'

1977 Stations: 8 Feb: 15° ♎ 11'; 17 May: 1° ♎ 27'

JAN ♎	FEB ♎	MAR ♎	APR ♎	MAY ♎	JUN ♎
10°36'	14°58'	13°41' **R**	7°29'	2°18'	2°17' **D**

JUL ♎	AUG ♎	SEP ♎	OCT ♏	NOV ♏	DEC ♐
7°13'	15°43'	26°23'	8°1'	20°53'	3°43'

1978 Stations: 24 May: 26° ♑ 59'; 2 Sept: 14° ♑ 23'

JAN	FEB	MAR	APR	MAY	JUN
♐	♐	♑	♑	♑	♑
16°57'	29°36'	9°57'	19°25'	25°27'	26°57' **R**

JUL	AUG	SEP	OCT	NOV	DEC
♑	♑	♑	♑	♑	≈
23°16'	17°12'	14°23'	16°56' D	24°4'	3°50'

1979 Stations: 9 Sept: 9° ♉ 52'; 14 Dec: 26° ♈12'

JAN	FEB	MAR	APR	MAY	JUN
≈	≈	♓	♓	♈	♈
15°45'	28°39'	10°37'	23°45'	5°59'	17°43'

JUL	AUG	SEP	OCT	NOV	DEC
♈	♉	♉	♉	♉	♈
27°39'	5°36'	9°43'	8°22' **R**	2°4'	26°46' **R**

1980 Stations: 19 Dec: 21° ♌ 42'

JAN	FEB	MAR	APR	MAY	JUN
♈	♉	♉	♉	♊	♊
27°5' D	3°3'	11°52'	23°12'	5°9'	17°55'

JUL	AUG	SEP	OCT	NOV	DEC
♋	♋	♋	♌	♌	♌
0°21'	13°1'	25°9'	5°56'	15°8'	20°40'

1981 Stations: 23 Mar: 7° ♌ 26'

JAN	FEB	MAR	APR	MAY	JUN
♌	♌	♌	♌	♌	♌
20°47' **R**	14°55'	9°2'	7°37' D	11°48'	19°58'

JUL	AUG	SEP	OCT	NOV	DEC
♍	♍	♍	♎	♎	♏
0°2'	11°45'	24°17'	6°51'	19°56'	2°19'

1982 Stations: 28 Mar: 2° ♐ 51'; 7 Jul: 19° ♏ 49'

JAN ♏ 14°11'	FEB ♏ 24°11'	MAR ♐ 0°29'	APR ♐ 2°49' **R**	MAY ♏ 29°34'	JUN ♏ 23°18'
JUL ♏ 19°57'	AUG ♏ 21°51'D	SEP ♏ 28°19'	OCT ♐ 7°30'	NOV ♐ 18°55'	DEC ♑ 1°7'

1983 Stations: 13 Jul: 15° ♓ 20'; 21 Oct: 2° ♓ 23'

JAN ♑ 14°22'	FEB ♑ 27°43'	MAR ♒ 9°26'	APR ♒ 21°34'	MAY ♓ 1°53'	JUN ♓ 10°11'
JUL ♓ 14°52'	AUG ♓ 14°36' **R**	SEP ♓ 9°18'	OCT ♓ 3°43'	NOV ♓ 2°43'D	DEC ♓ 7°13'

1984 Stations: 28 Oct: 27° ♓ 53'

JAN ♓ 15°32'	FEB ♓ 26°58'	MAR ♈ 8°34'	APR ♈ 21°31'	MAY ♉ 4°8'	JUN ♉ 16°55'
JUL ♉ 28°43'	AUG ♊ 9°57'	SEP ♊ 19°29'	OCT ♊ 25°55'	NOV ♊ 27°44' **R**	DEC ♊ 23°31' **R**

1985 Stations: 27 Jan: 13° ♊ 35'

JAN ♊ 16°26' **R**	FEB ♊ 13°37'D	MAR ♊ 16°33'	APR ♊ 24°9'	MAY ♋ 4°3'	JUN ♋ 15°42'
JUL ♋ 27°42'	AUG ♌ 10°9'	SEP ♌ 23°21'	OCT ♍ 5°37'	NOV ♍ 17°37'	DEC ♍ 27°52'

1986 Stations: 1 Feb: 9° ♎3′; 11 May: 25° ♍19′

JAN ♎	FEB ♎	MAR ♎	APR ♎	MAY ♍	JUN ♍
5°47′	9°3′ R	6°39′	0°2′	25°39′	26°47′D

JUL ♎	AUG ♎	SEP ♎	OCT ♏	NOV ♏	DEC ♐
2°31′	11°32′	22°31′	4°20′	17°17′	0°8′

1987 Stations: 17 May: 20° ♑49′; 26 Aug: 8° ♑1′

JAN ♐	FEB ♐	MAR ♑	APR ♑	MAY ♑	JUN ♑
13°17′	25°46′	5°50′	14°47′	20°0′	20°23′ R

JUL ♑	AUG ♑	SEP ♑	OCT ♑	NOV ♑	DEC ♑
15°50′	10°1′	8°19′D	11°53′	19°42′	29°52′

1988 Stations: 2 Sept: 3° ♉44′; 7 Dec: 20° ♈3′

JAN ♒	FEB ♒	MAR ♓	APR ♓	MAY ♈	JUN ♈
12°1′	25°2′	7°28′	20°33′	2°38′	14°5′

JUL ♈	AUG ♉	SEP ♉	OCT ♉	NOV ♈	DEC ♈
23°35′	0°48′	3°44′	1°6′ R	24°25′	20°11′ R

1989 Stations: 14 Dec: 15° ♌34′

JAN ♈	FEB ♈	MAR ♉	APR ♉	MAY ♊	JUN ♊
21°54′D	28°48′	7°48′	19°25′	1°30′	14°20′

JUL ♊	AUG ♋	SEP ♋	OCT ♌	NOV ♌	DEC ♌
26°46′	9°20′	21°19′	1°47′	10°26′	15°6′

1990 Stations: 17 Mar: 1° ♌ 17′

JAN ♌	FEB ♌	MAR ♌	APR ♌	MAY ♌	JUN ♌
13°59′ **R**	7°26′	2°9′	1°56′ **D**	7°2′	15°45′
JUL ♌	AUG ♍	SEP ♍	OCT ♎	NOV ♎	DEC ♎
26°8′	8°3′	20°41′	3°16′	16°18′	28°33′

1991 Stations: 24 Mar: 26° ♏ 47′; 1 Jul: 13° ♏ 41′

JAN ♏	FEB ♏	MAR ♏	APR ♏	MAY ♏	JUN ♏
10°8′	19°40′	25°15′	26°27′ **R**	22°12′	16°2′
JUL ♏	AUG ♏	SEP ♏	OCT ♐	NOV ♐	DEC ♐
13°41′ **D**	16°39′	23°51′	3°28′	15°8′	27°29′

1992 Stations: 7 Jul: 9° ♓ 11′; 15 Oct: 26° ♒ 14′

JAN ♑	FEB ♑	MAR ♒	APR ♒	MAY ♒	JUN ♓
10°47′	24°6′	6°8′	18°0′	27°54′	5°32′
JUL ♓	AUG ♓	SEP ♓	OCT ♒	NOV ♒	DEC ♓
9°7′	7°31′ **R**	1°55′	26°49′	27°13′ **D**	2°44′

1993 Stations: 21 Oct: 21° ♓ 44′

JAN ♓	FEB ♓	MAR ♈	APR ♈	MAY ♉	JUN ♉
12°3′	23°32′	4°54′	17°55′	0°3′	13°15′
JUL ♉	AUG ♊	SEP ♊	OCT ♊	NOV ♊	DEC ♊
24°55′	5°52′	14°55′	20°35′	21°14′ **R**	16°5′ **R**

1994 Stations: 24 Jan: 7° ♓ 42′

JAN ♓	FEB ♓	MAR ♓	APR ♊	MAY ♋	JUN ♋
9°20′ R	7°44′ D	11°36′	19°51′	0°7′	11°58′

JUL ♋	AUG ♌	SEP ♌	OCT ♍	NOV ♍	DEC ♍
24°5′	6°54′	19°44′	1°53′	13°40′	23°31′

1995 Stations: 29 Jan: 3° ♎ 11′; 4 May: 19° ♍ 10′

JAN ♎	FEB ♎	MAR ♍	APR ♍	MAY ♍	JUN ♍
0°43′	2°53′ R	29°26′	22°38′	19°14′	21°28′ D

JUL ♍	AUG ♎	SEP ♎	OCT ♏	NOV ♏	DEC ♏
27°57′	7°27′	18°43′	0°41′	13°42′	26°32′

1996 Stations: 11 May: 14° ♑ 41′; 19 Aug: 1° ♑ 53′

JAN ♐	FEB ♐	MAR ♑	APR ♑	MAY ♑	JUN ♑
9°36′	21°53′	1°58′	10°14′	14°25′	13°27′ R

JUL ♑	AUG ♑	SEP ♑	OCT ♑	NOV ♑	DEC ♑
8°8′	2°56′	2°34′ D	7°13′	15°44′	26°20′

1997 Stations: 26 Aug: 27° ♈ 33′; 1 Dec: 13° ♈ 56′

JAN ♒	FEB ♒	MAR ♓	APR ♓	MAY ♓	JUN ♈
8°44′	21°52′	3°53′	16°54′	28°51′	10°2′

JUL ♈	AUG ♈	SEP ♈	OCT ♈	NOV ♈	DEC ♈
19°7′	25°36′	27°29′ R	23°48′	17°6′	13°56′ D

1998 Stations: 7 Dec: 9° ♌ 26′

JAN ♈	FEB ♈	MAR ♉	APR ♉	MAY ♉	JUN ♊
16°47′	24°25′	3°48′	15°40′	27°53′	10°45′
JUL ♊	AUG ♋	SEP ♋	OCT ♋	NOV ♌	DEC ♌
23°9′	5°38′	17°25′	27°33′	5°36′	9°19′

1999 Stations: 11 Mar: 25° ♋ 9′

JAN ♌	FEB ♋	MAR ♋	APR ♋	MAY ♌	JUN ♌
6°59′ R	29°59′	25°29′	26°28′ D	2°22′	11°36′
JUL ♌	AUG ♍	SEP ♍	OCT ♍	NOV ♎	DEC ♎
22°17′	4°22′	17°5′	29°41′	12°39′	24°44′

2000 Stations: 17 Mar: 20° ♏ 39′; 23 Jun: 7° ♏ 18′

JAN ♏	FEB ♏	MAR ♏	APR ♏	MAY ♏	JUN ♏
6°2′	15°1′	19°55′	19°44′ R	14°32′	8°48′
JUL ♏	AUG ♏	SEP ♏	OCT ♏	NOV ♐	DEC ♐
7°43′ D	11°50′	19°47′	29°51′	11°48′	24°17′

2001 Stations: 29 Jun: 3° ♓ 40′; 7 Oct: 20° ♒ 10′

JAN ♑	FEB ♑	MAR ♒	APR ♒	MAY ♒	JUN ♓
7°40′	20°48′	2°19′	14°1′	23°37′	0°45′
JUL ♓	AUG ♓	SEP ♒	OCT ♒	NOV ♒	DEC ♒
3°38′ R	1°13′	25°4′	20°56′	22°11′ D	28°18′

2002 Stations: 14 Oct: 15° ♓ 56'

JAN ♓	FEB ♓	MAR ♈	APR ♈	MAY ♈	JUN ♉
7°59'	19°41'	1°9'	14°13'	26°50'	9°29'
JUL ♉	AUG ♊	SEP ♊	OCT ♊	NOV ♊	DEC ♊
21°1'	1°45'	10°25'	15°28'	15°16' **R**	9°33' **R**

2003 Stations: 22 Jan: 2° ♊ 15'

JAN ♊	FEB ♊	MAR ♊	APR ♊	MAY ♊	JUN ♋
3°12' **R**	2°30' D	7°0'	15°41'	26°11'	8°12'
JUL ♋	AUG ♌	SEP ♌	OCT ♌	NOV ♍	DEC ♍
20°22'	3°12'	16°0'	28°4'	9°39'	19°11'

II.
DARK SUNS

Table IIa — ☿ : Mean Inferior ☌ ☉: 16° ♊ 44′
Radius of Orbit: 0.16068 A.U. = 14,936,812 Miles

☉	☿		☉	☿		☉	☿
0° ♈	20° ♓ 48′		15° ♌	23° ♌ 27′		16° ♐ 44′	16° ♐ 44′
15° ♈	6° ♈ 20′		0° ♍	9° ♍ 8′		0° ♑	28° ♐ 10′
0° ♉	22° ♈ 33′		*7° ♍ 44′	16° ♍ 59′		15° ♑	11° ♑ 12′
15° ♉	9° ♉ 26′		15° ♍	24° ♍ 8′		0° ♒	24° ♑ 22′
0° ♊	26° ♉ 54′		0° ♎	8° ♎ 33′		15° ♒	7° ♒ 50′
15° ♊	14° ♊ 40′		15° ♎	22° ♎ 28′		0° ♓	21° ♒ 50′
16° ♊ 44′	16° ♊ 44′		0° ♏	6° ♏ 0′		15° ♓	5° ♓ 57′
0° ♋	2° ♋ 29′		15° ♏	19° ♏ 14′		*26° ♓ 14′	16° ♓ 59
15° ♋	20° ♋ 3′		0° ♐	2° ♐ 18′			
0° ♌	7° ♌ 4′		15° ♐	15° ♐ 14′			

*Maximum Arc from ☉: 9°15′

Table IIb — ♀ : Mean Inferior ☌ ☉: 11° ♌ 16′
Radius of Orbit: 0.009792 A.U. = 910,264 Miles

☉	♀		☉	♀		☉	♀
0° ♈	29° ♓ 35′		11° ♌ 16′	11° ♌ 16′		0° ♑	0° ♑ 22′
0° ♉	29° ♈ 28′		0° ♍	0° ♍ 11′		0° ♒	0° ♒ 7′
*11° ♉ 50′	11° ♉ 16′		0° ♎	0° ♎ 25′		11° ♒ 16′	11° ♒ 16′
0° ♊	29° ♉ 28′		0° ♏	0° ♏ 32′		0° ♓	29° ♒ 49′
0° ♋	29° ♊ 38′		*10° ♏ 42′	11° ♏ 16′			
0° ♌	29° ♋ 53′		0° ♐	0° ♐ 31′			

*Maximum Arc from ☉: 0°34′

Table IIc — ● : Mean Inferior ☌☉: 11° ♋45'
Radius of Orbit: 0.0334 A.U. = 3,104,864 Miles

☉	●	☉	●	☉	●
0° ♈	28° ♓9'	0° ♌	0° ♌37'	0° ♑	0° ♑23'
*13° ♈40'	11° ♈45'	0° ♍	1° ♍27'	11° ♑45'	11° ♑45'
0° ♉	28° ♈10'	0° ♎	1° ♎53'	0° ♒	29° ♑25'
0° ♊	28° ♉42'	*9° ♎50'	11° ♎45'	0° ♓	28° ♒37'
0° ♋	29° ♊53'	0° ♏	1° ♏47'		
11° ♋45'	11° ♋45'	0° ♐	1° ♐14'		

*Maximum Arc from ☉: 1°55'

Table IId — ☾ : Mean Inferior ☌☉: 5° ♓12'
Radius of Orbit: 0.28272 A.U. = 26,281,651 Miles

☉	☾	☉	☾	☉	☾
0° ♈	8° ♈55'	15° ♌	19° ♌24'	*21° ♐37'	5° ♐12'
15° ♈	27° ♈51'	0° ♍	1° ♍8'	0° ♑	14° ♐0'
0° ♉	15° ♉10'	5° ♍12'	5° ♍12'	15° ♑	0° ♑22'
15° ♉	1° ♓11'	15° ♍	12° ♍51'	0° ♒	18° ♑13'
*18° ♉47'	5° ♓12'	0° ♎	24° ♍38'	15° ♒	7° ♒34'
0° ♊	16° ♊0'	15° ♎	6° ♎34'	0° ♓	26° ♒15'
15° ♊	29° ♊48'	0° ♏	18° ♎49'	5° ♓12'	5° ♓12'
0° ♋	12° ♋51'	15° ♏	1° ♏28'	15° ♓	18° ♓45'
15° ♋	25° ♋23'	0° ♐	14° ♏44'		
0° ♌	7° ♌31'	15° ♐	28° ♏48'		

*Maximum Arc from ☉: 16°25'

Table IIe — ♃ : Mean Inferior ☌☉: 13° ♈54'
Radius of Orbit: 0.4992 A.U. = 46,405,632 Miles

☉	♃	☉	♃	☉	♃
0° ♈	17° ♓1'	5° ♉	23° ♉35'	15° ♋	11° ♌17'
5° ♈	26° ♓20'	10° ♉	1° ♓42'	0° ♌	22° ♌49'
10° ♈	6° ♈8'	15° ♉	9° ♓5'	15° ♌	3° ♍45'
13° ♈54'	13° ♈54'	20° ♉	16° ♓14'	0° ♍	14° ♍16'
16° ♈	16° ♈2'	0° ♊	28° ♓43'	15° ♍	24° ♍31'
20° ♈	26° ♈1'	*13° ♊54'	13° ♋51'	0° ♎	4° ♎37'
25° ♈	5° ♉40'	15° ♊	14° ♋53'	13° ♎54'	13° ♎54'
0° ♉	14° ♉54'	0° ♋	28° ♋47'	15° ♎	14° ♎38'

Table IIe — •+ : Mean Inferior ☌ ☉: 13° ♈ 54′
Radius of Orbit: 0.4992 A.U. = 46,405,632 Miles

☉	•+		☉	•+		☉	•+
0° ♏	24° ♎ 39′		15° ♑	18° ♐ 17′		5° ♓	7° ♒ 52′
15° ♏	4° ♏ 46′		0° ♒	0° ♑ 58′		10° ♓	14° ♒ 34′
0° ♐	15° ♏ 3′		*13 ♒ 48′	13° ♑ 51′		15° ♓	21° ♒ 48′
15° ♐	25° ♏ 37′		15° ♒	15° ♑ 9′		20° ♓	29° ♒ 36′
0° ♑	6° ♐ 38′		0° ♓	1° ♒ 44′		25° ♓	7° ♓ 58′

*Maximum Arc from ☉: 29°57′

Table IIf — ƫ : Mean (Superior) ☌ ☉: 1° ♑ 30′
Radius of Orbit: 1.06848 A.U. = 99,325,900 Miles

☉	ƫ		☉	ƫ		☉	ƫ
0° ♈	13° ♒ 54′		0° ♋	22° ♑ 2′		0° ♍	4° ♏ 10′
15° ♈	20° ♒ 50′		1° ♋	8° ♑ 45′		15° ♍	10° ♏ 48′
0° ♉	27° ♒ 33′		1° ♋ 30′	1° ♑ 30′		0° ♎	17° ♏ 2′
15° ♉	3° ♓ 49′		2° ♋	24° ♐ 15′		15° ♎	24° ♏ 46′
0° ♊	8° ♓ 57′		3° ♋	10° ♐ 58′		0° ♏	1° ♐ 53′
5° ♊	10° ♓ 3′		4° ♋	29° ♏ 41′		15° ♏	9° ♐ 4′
10° ♊	10° ♓ 46′	℞ ↓	5° ♋	20° ♏ 49′		0° ♐	16° ♐ 17′
15° ♊	10° ♓ 5′		6° ♋	14° ♏ 4′		15° ♐	23° ♐ 31′
20° ♊	7° ♓ 27′		7° ♋	8° ♏ 58′		0° ♑	0° ♑ 46′
22° ♊	4° ♓ 52′		9° ♋	2° ♏ 19′		1° ♑ 30′	1° ♑ 30′
24° ♊	0° ♓ 41′		12° ♋	26° ♎ 40′		15° ♑	8° ♑ 2′
26° ♊	24° ♒ 2′		15° ♋	24° ♎ 3′		0° ♒	15° ♑ 16′
27° ♊	18° ♒ 56′		22° ♋	22° ♎ 33′	D ↓	15° ♒	22° ♑ 30′
28° ♊	12° ♒ 11′		0° ♌	23° ♎ 17′		0° ♓	29° ♑ 41′
29° ♊	3° ♒ 19′		15° ♌	28° ♎ 4′		15° ♓	6° ♒ 49′

NB. ℞ 6 weeks: 10° ♊ - 22° ♋.

Table IIg — ⚷ : Mean (Superior) ☌ ☉: 7° ♈ 40′
Radius of Orbit: 10.37063 A.U. = 964,054,000 Miles

☉	⚷		☉	⚷		☉	⚷	
0° ♈	7° ♈ 0′		15° ♉	10° ♈ 45′		15° ♋	13° ♈ 9′	
7° ♈ 40′	7° ♈ 40′		0° ♊	11° ♈ 40′		0° ♌	12° ♈ 54′	℞ ↓
15° ♈	8° ♈ 18′		15° ♊	12° ♈ 32′		15° ♌	12° ♈ 17′	
0° ♉	9° ♈ 34′		0° ♊	13° ♈ 1′		0° ♍	11° ♈ 17′	

Table IIg — ⚷ : Mean (Superior) ☌ ☉: 7° ♈ 40′
Radius of Orbit: 10.37063 A.U. = 964,054,000 Miles

☉	⚷		☉	⚷		☉	⚷
15° ♍	9° ♈ 59′		15° ♏	4° ♈ 5′		0° ♒	2° ♈ 47′
0° ♎	8° ♈ 29′		0° ♐	3° ♈ 5′		15° ♒	3° ♈ 33′
7° ♎ 40′	7° ♈ 40′		15° ♐	2° ♈ 26′		0° ♓	4° ♈ 34′
15° ♎	6° ♈ 54′		0° ♑	2° ♈ 12′	D	15° ♓	5° ♈ 44′
0° ♏	5° ♈ 23′		15° ♑	2° ♈ 19′	↓	0° ♈	7° ♈ 0′

NB. 5 months ℞: ☉ ♌ - ☉ ♑.

Table IIh — ♅ : Mean (Superior) ☌ ☉: 21° ♓ 35′
Radius of Orbit: 1.8048 A.U. = 167,774,000 Miles

☉	♅		☉	♅		☉	♅	
0° ♈	24° ♓ 35′		0° ♍	14° ♈ 30′		25° ♎	22° ♒ 4′	
15° ♈	29° ♓ 52′		5° ♍	10° ♈ 8′		0° ♏	20° ♒ 19′	
8° ♉	5° ♈ 3′		10° ♍	5° ♈ 15′		15° ♏	18° ♒ 3′	D
15° ♉	10° ♈ 2′		15° ♍	29° ♓ 37′		0° ♐	18° ♒ 54′	↓
0° ♊	14° ♈ 42′		20° ♍	23° ♓ 33′		15° ♐	21° ♒ 16′	
15° ♊	18° ♈ 52′		21° ♍ 35′	21° ♓ 35′		0° ♑	24° ♒ 47′	
0° ♋	22° ♈ 16′		25° ♍	17° ♓ 22′		15° ♑	29° ♒ 1′	
15° ♋	24° ♈ 28′		0° ♎	11° ♓ 26′		0° ♒	3° ♓ 43′	
0° ♌	24° ♈ 43′	℞	5° ♎	6° ♓ 2′		15° ♒	8° ♓ 44′	
15° ♌	21° ♈ 42′	↓	10° ♎	1° ♓ 25′		0° ♓	13° ♓ 56′	
20° ♌	20° ♈ 18′		15° ♎	27° ♒ 33′		15° ♓	19° ♓ 15′	
25° ♌	17° ♈ 40′		20° ♎	24° ♒ 27′		21° ♓ 35′	21° ♓ 35′	

NB. 3 months ℞ while close to Earth and movement greater.

Table IIi — ♇ : Mean (Superior) ☌ ☉: 13° ♉ 0′
Radius of Orbit: 19.7 A.U. = 1,831.312 Million Miles

☉	♇		☉	♇		☉	♇
0° ♈	11° ♉ 6′		0° ♋	15° ♉ 3′		15° ♎	14° ♉ 19′
15° ♈	11° ♉ 41′		15° ♋	15° ♉ 30′		0° ♏	13° ♉ 1′
0° ♉	12° ♉ 23′		0° ♌	15° ♉ 47′		13° ♏	13° ♉ 0′
13° ♉	13° ♉ 0′		15° ♌	15° ♉ 54′	℞	15° ♏	12° ♉ 59′
15° ♉	13° ♉ 6′		0° ♍	15° ♉ 49′	↓	0° ♐	12° ♉ 47′
0° ♊	13° ♉ 48′		15° ♍	15° ♉ 31′		15° ♐	11° ♉ 24′
15° ♊	14° ♉ 29′		0° ♎	15° ♉ 3′		0° ♑	10° ♉ 48′

Table IIi — ♀ : Mean (Superior) ☌☉: 13° ♅ 0'
            Radius of Orbit: 19.7 A.U. = 1,831.312 Million Miles

☉	♀	☉	♀		☉	♀
15° ♉	10° ♅ 23'	15° ♒	10° ♅ 6'	D	15° ♓	10° ♅ 37'
0° ♒	10° ♅ 8'	0° ♓	10° ♅ 17'		0° ♈	11° ♅ 6'

NB. 6 months ℞ 15° ♌ - 15° ♒.

Table IIj & k — ♆ : Mean Inferior ☌☉, see Table k
            Mean Radius of Orbit: 0.517032 A.U. = 48,063,294 Miles
♆ varies its position by far too much to permit a mean figure for the century.
Table k gives mean ☌☉ for each year. Table j requires the user to calculate
how far ☉ is past (after) inferior ☌ for the year in question. The second
column in each pair then shows how far ♆ is past the same ☌ position,
by absolute longitude.

☉ Past ☌ By	♆ Past ☌ By	☉ Past ☌ By	♆ Past ☌ By	☉ Past ☌ By	♆ Past ☌ By
0°	0°	110°	132°22'	250°	227°38'
5°	9°59'	120°	139°35'	260°	235°03'
10°	19°59'	130°	146°31'	270°	242°40'
15°	29°58'	140°	153°22'	280°	250°54'
20°	38°21'	150°	160°07'	290°	259°39'
25°	46°43'	160°	166°46'	300°	268°53'
30°	55°06'	170°	173°24'	*310°	279°40'
40°	68°20'	180°	180°00'	320°	291°40'
50°	80°20'	190°	186°36'	330°	304°54'
*60°	91°07'	200°	193°14'	335°	313°17'
70°	100°21'	210°	199°53'	340°	321°39'
80°	109°05'	220°	206°38'	345°	330°02'
90°	117°20'	230°	213°29'	350°	340°01'
100°	124°57'	240°	220°25'	355°	350°00'

*Maximum Arc from ☉: 31°8' when ☉ is 58°52' or 301°8' Past Inferior ☌.

Table k — ♆ : Mid-year positions of Inferior ☌☉

1901	1902	1903	1904	1905	1906	1907	1908
13° ♈35'	18° ♈24'	28° ♈47'	10° ♅46'	16° ♅54'	29° ♅55'	5° ♓21'	7° ♓33'

1909	1910	1911	1912	1913	1914	1915	1916
5° ♓42'	28° ♅32'	22° ♅30'	3° ♅44'	26° ♈51'	28° ♈7'	6° ♅5'	17° ♅55'

Table k — ♆ : Mid-year positions of Inferior ☌ ☉

1917	1918	1919	1920	1921	1922	1923	1924
0° ♊ 24′	11° ♊ 3′	18° ♊ 40′	22° ♊ 37′	21° ♊ 46′	13° ♊ 43′	27° ♉ 0′	7° ♉ 28′
1925	1926	1927	1928	1929	1930	1931	1932
24° ♈ 57′	22° ♈ 16′	26° ♈ 57′	6° ♉ 38′	19° ♉ 8′	1° ♊ 46′	11° ♊ 57′	17° ♊ 34′
1933	1934	1935	1936	1937	1938	1939	1940
16° ♊ 33′	6° ♊ 41′	19° ♉ 43′	3° ♉ 54′	25° ♈ 43′	24° ♈ 45′	28° ♈ 46′	6° ♉ 25′
1941	1942	1943	1944	1945	1946	1947	1948
16° ♉ 53′	29° ♉ 14′	11° ♊ 54′	22° ♊ 38′	28° ♊ 28′	25° ♊ 12′	9° ♊ 48′	19° ♉ 43′
1949	1950	1951	1952	1953	1954	1955	1956
7° ♉ 14′	4° ♉ 17′	6° ♉ 52′	12° ♉ 29′	20° ♉ 3′	29° ♉ 3′	8° ♊ 51′	17° ♊ 57′
1957	1958	1959	1960	1961	1962	1963	1964
21° ♊ 33′	8° ♊ 2′	11° ♉ 1′	24° ♈ 42′	24° ♈ 21′	29° ♈ 24′	5° ♉ 45′	12° ♉ 8′
1965	1966	1967	1968	1969	1970	1971	1972
18° ♉ 2	23° ♉ 0′	26° ♉ 23′	26° ♉ 40′	20° ♉ 57′	7° ♉ 50′	24° ♈ 22′	19° ♈ 49′
1973	1974	1975	1976	1977	1978	1979	1980
24° ♈ 10′	2° ♉ 34′	11° ♉ 33′	19° ♉ 33′	25° ♉ 52′	0° ♊ 5′	1° ♊ 14′	26° ♉ 41′
1981	1982	1983	1984	1985	1986	1987	1988
11° ♉ 15′	16° ♈ 33′	27° ♓ 54′	24° ♓ 20′	1° ♈ 28′	14° ♈ 20′	28° ♈ 54′	11° ♉ 51′
1989	1990	1991	1992	1993	1994	1995	1996
21° ♉ 15′	26° ♉ 18′	26° ♉ 10′	19° ♉ 10′	6° ♉ 0′	17° ♈ 26′	5° ♈ 48′	2° ♈ 47′
1997	1998	1999	2000	2001			
7° ♈ 1′	16° ♈ 40′	29° ♈ 53′	12° ♉ 58′	21° ♉ 0′			

III.

PLANETARY NODES

Geocentric Mean Node Positions 1900-2000

Month	MERCURY		VENUS*		MARS*	
1st of:	North	South	North	South	North	South
JAN	27° ♉17'	24° ♐2'	20° ♒10'	29° ♐59'	6° ♈26'	8° ♐32'
FEB	29° ♒38'	17° ♉53'	26° ♓30'	18° ♉41'	12° ♈36'	19° ♐22'
MAR	24° ♓56'	12° ♒52'	17° ♈49'	6° ♒32'	22° ♈9'	26° ♐51'
APR	19° ♈41'	17° ♓21'	7° ♉42'	27° ♒55'	3° ♉56'	27° ♐24'
MAY	12° ♉10'	3° ♉55'	25° ♉19'	24° ♓59'	15° ♉39'	4° ♐2'
JUN	4° ♊52'	27° ♉6'	12° ♊47'	25° ♉14'	27° ♉43'	19° ♎51'
JUL	27° ♊11'	5° ♌35'	29° ♊28'	18° ♌18'	9° ♊2'	9° ♎38'
AUG	21° ♋47'	4° ♍37'	16° ♌58'	24° ♍0'	19° ♊54'	13° ♎50'
SEP	19° ♌38'	28° ♍40'	5° ♌22'	16° ♎42'	28° ♊46'	22° ♎45'
OCT	22° ♍11'	19° ♒58'	25° ♌19'	5° ♏34'	1° ♋47'	3° ♏13'
NOV	3° ♏47'	11° ♏18'	22° ♍35'	24° ♏6'	10° ♊28'	14° ♏59'
DEC	17° ♐37'	2° ♐3'	19° ♏27'	11° ♐46'	17° ♈4'	26° ♏44'

1st of:	JUPITER		SATURN		CHIRON	
JAN	9° ♋59'	9° ♉57'	24° ♋52'	22° ♉2'	5° ♏7'	26° ♈11'
FEB	3° ♋5'	14° ♉53'	21° ♋1'	24° ♉52'	5° ♏44'	26° ♈10'
MAR	29° ♓27'	18° ♉36'	18° ♋3'	27° ♉9'	4° ♏30'	26° ♈53'
APR	29° ♓4'	20° ♉51'	16° ♋54'	28° ♉41'	1° ♏25'	28° ♈13'
MAY	1° ♋21'	20° ♉20'	17° ♋24'	28° ♉50'	27° ♎38'	29° ♈42'
JUN	5° ♋17'	16° ♉28'	19° ♋15'	27° ♉25'	24° ♎21'	1° ♉4'
JUL	9° ♋48'	10° ♉13'	21° ♋49'	24° ♉47'	22° ♎41'	1° ♉57'

1st of:	JUPITER		SATURN		CHIRON	
AUG	14°♋29'	3°♑40'	24°♋45'	21°♑34'	22°♎45'	2°♉11'
SEP	18°♋28'	29°♐41'	27°♋25'	18°♑53'	24°♎21'	1°♉35'
OCT	20°♋48'	29°♐3'	29°♋9'	17°♑35'	26°♎56'	0°♉19'
NOV	20°♋34'	1°♑11'	29°♋31'	17°♑52'	0°♏5'	28°♈36'
DEC	16°♋51'	5°♑4'	28°♋4'	19°♑29'	2°♏59'	27°♈6'

1st of:	URANUS		NEPTUNE		PLUTO	
JAN	12°♓5'	14°♐47'	11°♌55'	9°♒55'	19°♋50'	19°♉19'
FEB	10°♓53'	15°♐59'	10°♌53'	10°♒54'	19°♋2'	20°♉8'
MAR	10°♓31'	16°♐29'	9°♌55'	11°♒49'	18°♋29'	20°♉44'
APR	10°♓56'	16°♐13'	9°♌12'	12°♒32'	18°♋11'	21°♉5'
MAY	11°♓57'	15°♐12'	8°♌59'	12°♒48'	18°♋17'	21°♉1'
JUN	13°♓20'	13°♐40'	9°♌15'	12°♒35'	18°♋42'	20°♉33'
JUL	14°♓44'	12°♐8'	9°♌55'	11°♒56'	19°♋19'	19°♉53'
AUG	15°♓52'	10°♐58'	10°♌49'	10°♒58'	20°♋2'	19°♉4'
SEP	16°♓27'	10°♐30'	11°♌45'	9°♒58'	20°♋37'	18°♉24'
OCT	16°♓16'	10°♐49'	12°♌28'	9°♒15'	20°♋57'	18°♉5'
NOV	15°♓17'	11°♐51'	12°♌48'	8°♒59'	20°♋55'	18°♉9'
DEC	13°♓45'	13°♐15'	12°♌36'	9°♒14'	20°♋31'	18°♉36'

* Periods of Rapid Movement, Venus and Mars

VENUS			VENUS			MARS		
Date	North	South	Date	North	South	Date	North	South
May 8th	29°♈18'	3°♈42'	Nov 8th	1°♎11'	28°♏4'	May 1st	15°♈39'	4°♐2'
15th	3°♓15'	14°♈24'	15th	11°♎42'	2°♐22'	8th	18°♈23'	22°♏52'
22nd	7°♓11'	27°♈51'	22nd	25°♎19'	6°♐29'	15th	21°♉8'	11°♏5'
29th	11°♓7'	15°♈55'	29th	13°♏19'	10°♐36'	22nd	23°♈51'	0°♐36'
Jun 1st	12°♓47'	25°♈14'	Dec 1st	19°♏27'	11°♐46'	29th	26°♈34'	22°♎28'
8th	16°♓40'	18°♓16'	8th	13°♐43'	15°♐55'	Nov 1st	10°♓28'	14°♏59'
15th	20°♓34'	11°♋7'	15th	8°♉15'	20°♐4'	8th	27°♈31'	17°♏43'
22nd	24°♓27'	0°♌10'	22nd	28°♉59'	24°♐12'	15th	12°♈40'	20°♏28'
29th	28°♓11'	14°♌47'	29th	14°♒48'	28°♐22'	22nd	29°♈13'	23°♏1'
Jul 1st	29°♓28'	18°♌18'	Jan 1st	20°♒10'	29°♐59'	29th	19°♈15'	25°♏57'
8th	3°♋23'	28°♌56'	8th	0°♓59'	4°♉10'			
15th	7°♋18'	7°♍36'	15th	4°♓47'	8°♉22'			

APPENDICES

A.
BODE'S LAW

This empirical formula for the distances of planets from the Sun was first propounded in 1766 by J. D. Titius, and later developed by J. E. Bode. The formula states that the mean distances of the planets from the Sun (in astronomical units) are equivalent to $0.4 + 0.3 \times 2^n$, where n = minus infinity, 0, 1, 2, and so on.

The results given by this are, for the most part, a very good match for the actual distances of the planets from the Sun. There are two anomalies in the system:

(a) Where $n = 3$, the result is 2.8, somewhere between Mars and Jupiter. There is no planet currently in this position, though the asteroids are in this region. Ceres' distance is 2.77.

(b) When $n = 6$ the result corresponds to Uranus's position, but when $n = 7$ the result corresponds to Pluto's; Neptune seems to have no place in the arrangement, unless Pluto has somehow taken Neptune's orbit, and should itself be elsewhere, or be a satellite of another planet.

Planet	n	Mean radius of orbit	Predicted radius by Bode
☿	$-\infty$	0.39	0.4
♀	0	0.72	0.7
⊕	1	1.0	1.0
♂	2	1.52	1.6
Ceres*	3	2.8	2.8
♃	4	5.2	5.2
♄	5	9.54	10.0

* 2.8 is the approximate figure for all the main belt of asteroids. The figure for ♁ is 2.7378.

Planet	n	Mean radius of orbit	Predicted radius by Bode
♅	6	19.2	19.6
♆ ⧸ ♇ ⧹	7	{ 30.1 } { 39.5 }	38.8

The next result, for n = 8, gives a predicted radius of 77.2 — very close to the position of the very recently confirmed outermost planet 'Transpluto' or 'Isis.

B.

PHAETHON, LUNA AND LILITH

Raymond Henry

In earlier times, there were two suns in the sky: the one we still have, and Phaethon. There were more moons, too: our original satellite was Lilith, and we did not acquire Luna until later, after the catastrophe that caused the fragmentation of Phaethon.

Confusion between Lilith, Luna, and Phaethon exceeds any other astronomical confusion known to us. Try, for example, this quote from the thirteenth century Kabbalistic volume, *The Zohar*:

> 'God made two great lights, The two lights ascended with the same dignity. The Moon, however, was not at ease with the Sun, and in fact each felt mortified by the other.'

The Zohar then quotes the 'Song of Songs' as Sun and Moon begin to dispute with each other, Sun saying contemptuously: 'How can a little candle shine at midday?'

'God then said to her (Moon): "Go and diminish thyself." '

And when Moon protested at this, God is quoted as adding:

' "Go thy way forth in the footsteps of the flock." From that time she has had no light of her own but derives her light from the Sun.'

The Zohar is talking about Lilith, not Luna, when it speaks of the Moon. Yet the passage above says quite plainly that the two lights at first had equal dignity (or power), and that only later, after being banished to head the flock (become chief planet, from our Earth-centred viewpoint) did she cease to have light of her own making. Lilith is too small ever to have had light of her own; she is even smaller than Luna, and would need to be at least ten times as large as Jupiter to become a star and be radiant. Nor could she, at her present mass, have been in conflict with the Sun: the Sun would scarcely notice her existence. What was in conflict with the Sun, and did have equal dignity with it, was Phaethon. This not-so-little candle certainly did

shine at midday, until its destruction; now its remnants, with cold
Ceres as *leader of the flock of the asteroids*, have no light of their own.

It is difficult to untangle Lilith from her mythology. She was Adam's
first wife, and she was unfaithful to him (with whom?). In some tales
it was she who gave him the apple, and in others she flatly refused
to let him breed with her. Whatever the astronomy behind the
mythology, she gets a very bad press. There may be a reason for this:
she may have collided with the Earth (if estimates of her size are correct,
and if this did ever happen, we are lucky still to be here). Or she may
have simply deserted us in the chaos of Phaethon's destruction. Where
is she now? Pluto is probably too large to have been Lilith. Some recent
notions have suggested Chiron; there is an interesting parallel in that
both bodies are held to represent a despised and rejected person who
had much wisdom, but efforts to identify the two bodies must be
thwarted by lack of knowledge about our earlier satellite. Followers
of Alice Bailey might suggest Vulcan, the esoteric inner planet: it is
our opinion that the point of significance attributed to Vulcan is in
fact the Dark Sun of Mercury, as explained in Part Two of this book.

Luna's arrival is well-documented. Tibetan accounts relate how we
first had one satellite, then two, then only one again — the new one,
during which events the Earth stood still, toppled over pole for pole,
so that the Sun stood still for a day and the satellite(s) went the other
way; then we tilted again, with whole oceans vanishing, new mountains
rising, and with Tibet itself 'raised up to the heavens'. It is still at high
altitude, of course, complete with a dried-up sea bed containing the
remains of what was once tropical fauna.

Judaic lore contains the same tale, though somewhat abridged. In
Genesis we have the story of the Atlantis disaster:

'The same day were all the fountains of the great deep broken up,
and all the windows of heaven were opened.'

We are also told that this happened on the seventeenth day of the
second month, which, remembering that the Judaic New Year starts
at the Autumn Equinox, coincides with the Phaethon disaster
occurring in our present period of Scorpio, with effects felt down here
at about 18 or 19 degrees of the same. Is this the reason why the
eighteenth degree of Scorpio is usually regarded as disastrous by
astrologers?

(How infuriating this is! We have a day and month for the disaster,
and zodiacal positions to support it, but no year! Dr Kamenenko's
estimate is about 7500 BC.)

Luna is an older body than the Earth, as has been established by the Apollo missions. It could well have belonged to the older of the two stars, which would have been Phaethon, and it came to us during the disaster. Whether this was by accident or design depends on your imagination, but there are more than enough mythological references to gods and demigods who live there to promote pondering.

It is worth considering life on Earth in the era of two suns. There would have been nothing to match the darkness of our winter nights; at most we would have had a brief twilight twice each day. A few of the very brightest stars would have been just visible in those twilights — 'between the lights', literally — and Sirius would have ranked as the most visible of them all, as 'chief amongst the stars', which is what its name means. The planets would have been seen either in the twilight zones, or in rather partial eclipse transits across either of the suns.

After the disaster, and after some 1500 years or more, when the sky cleared enough to see anything at all, it would indeed have been:

'A new Heaven and a new Earth, for the old Heaven and the old Earth had passed away, and were no more' (Revelations).

Only one Sun now remained, to be given the name Solus, 'alone', and half the day was now in darkness. For the first time the remaining creatures on Earth were aware of the binary cycle of night and day, and may have developed intellect and psyche in response to this. At night they were able to see for the first time too how large was the universe they lived in, and how many millions of stars there were. It may be that the monotheistic religions have their origins here, in contemplation of the infinite and unknowable expanse of the night sky. In time, we noticed that one or two of the stars were variable in their light: binaries. The names given to these stars, like Algol, the Demon, show how much we distrust them, and how we have learned from our past.

INDEX

Schliemann, Heinrich, 28-9, 34-5
54
Schumacher, Ernst, 28, 34, 54
Sirius, 170
Skinner, B. F., 72, 83-4
Smith, Joseph, 41, 54
Sun Yat-sen, 106

Tchaikovsky, Petr, 104
Teilhard de Chardin, Pierre, 40,
54
Tektites, 15
Teresa of Calcutta, 85-7
Tesla, Nikola, 44, 54
Thatcher, Margaret, 35, 43, 54
Truman, Harry, 106

Velikovsky, Immanuel, 17
Victoria, 38
Vindemiatrix, 65
Vulcan, 72, 169

Wall St crash, 34
Watson, James, 48-51
Weisenthal, Simon, 40, 45, 46-7,
48, 74
Wheeler, Elsie, 116-8
Wilkins, Maurice, 32, 48-51

Yeats, W. B., 45, 54
Young, Brigham, 41, 54

Of further interest . . .

SOLAR AND LUNAR RETURNS
by John Filbey

The solar return is a 'revolution' chart erected for the exact time when the sun returns to the same longitude as at your birth. This occurs each year within a day or two of your birthday. Similarly, the lunar return is calculated for the moon, which recurs every 27 days or so.

Unlike other types of return, these are personal and calculable separately for the individual. The solar chart indicates events which will be experienced during the year from the date of the return, while the lunar chart gives a more short-term indication of conditions.

This book is a comprehensive study of the various techniques of solar and lunar returns and gives detailed instructions in their calculation and application. The author discusses the tropical and sidereal zodiacs, explaining their origins and relative merits, and the phenomenon of the Precession of the Equinoxes in detail. A mass of worked examples backs up the theoretical side of the work, clarifying the several techniques involved.

John Filbey is a former tutor in computation and method for the Faculty of Astrological Studies and was also the Chairman of its examining board. In 1977 he was awarded the Fellowship of the Faculty in recognition of his services. He is also a member of the Astrological Association and author of Natal Charting, The Astrologer's Companion *and* Astronomy for Astrologers.